# THE WHOLE TRUTH

# The Whole Truth

## A COSMOLOGIST'S REFLECTIONS ON THE SEARCH FOR OBJECTIVE REALITY

*P. J. E. Peebles*

PRINCETON UNIVERSITY PRESS
PRINCETON & OXFORD

Published by Princeton University Press
41 William Street, Princeton, New Jersey 08540
99 Banbury Road, Oxford OX2 6JX

press.princeton.edu

Library of Congress Cataloging-in-Publication Data

Names: Peebles, P. J. E. (Phillip James Edwin), author.
Title: The whole truth : a cosmologist's reflections on the search for objective reality /
   P. J. E. Peebles.
Description: Princeton : Princeton University Press, [2022] | Includes bibliographical
   references and index.
Identifiers: LCCN 2021051638 (print) | LCCN 2021051639 (ebook) |
   ISBN 9780691231358 (hardback) | ISBN 9780691231365 (ebook)
Subjects: LCSH: Science—Philosophy. | Physics. | Cosmology. | Reality. |
   BISAC: SCIENCE / Space Science / Cosmology | SCIENCE / History
Classification: LCC Q175 .P372 2022 (print) | LCC Q175 (ebook) |
   DDC 501—dc23/eng/20220128
LC record available at https://lccn.loc.gov/2021051638
LC ebook record available at https://lccn.loc.gov/2021051639

British Library Cataloging-in-Publication Data is available

Editorial: Ingrid Gnerlich and Whitney Rauenhorst
Production Editorial: Mark Bellis
Jacket Design: Chris Ferrante
Production: Danielle Amatucci
Publicity: Matthew Taylor and Kate Farquhar-Thomson
Copyeditor: Bhisham Bherwani

Jacket Credit: Galaxy orbits flowing out of voids and impinging on regions of high density. Constructed by Edward Shaya (University of Maryland), Brent Tully (University of Hawaii), Daniel Pomarede (University of Paris-Saclay), and Yehuda Hoffman (Hebrew University, Jerusalem)

This book has been composed in Miller

Printed on acid-free paper. ∞

Printed in the United States of America

10 9 8 7 6 5 4 3 2 1

To Alison, my best friend

CONTENTS

What is at the core of what we do in natural science? I say that we are searching for the nature of reality, but what does that mean? You will not find ready answers from most working scientists; they would rather work on the problems arising in the research at hand. But there are lines of thought drawn from the philosophy, sociology, and history of science that are good descriptions of what I see is at the heart of what we are doing in science. I mean to explain, with illustrations from cosmology, the theory of the expansion of the universe from a hot dense state.

I began to work on cosmology a half century ago, when ideas were speculative and evidence scant, and I have seen it grow into one of the best-established of the branches of physical science. The experience has led me to reflect on what my colleagues and I have been up to, and what we have learned, apart from how to work out more or less well posed problems as they arise. The result is presented in this book.

I mean this book to be accessible to those who are not familiar with the language of physics but are interested in what scientists are doing and why. Colleagues in science may be surprised to find me discussing aspects of sociology and philosophy, and the evidence from natural science for objective physical reality. These are not familiar topics of conversation in my crowd, but I think they are essential for a more complete picture of what we are learning from research in natural science. I am not attempting an assessment of thinking about reality in the philosophy and sociology of science, but I hope authorities in those fields will find it interesting to see how some ideas drawn from their disciplines find resonance with the practice of natural science.

Scientists have different ways of thinking about our subject. I am on the empiricist side of the spectrum; I love to see debates settled by measurements. But despite my empiricist inclination to confine attention to the state of the experimental or observational tests of physical theories, and leave it at that, I have been led to think that curiosity-driven science has produced good evidence of an abstract notion: objective reality. The notion as it is applied in science can be falsified, by failure of the scientific method. It can never be proved, because the accuracy of empirical evidence always is limited. The theme of this book is that the empirical results from natural science, presented in the worked example of physical cosmology, add up to a persuasive case for observer independent reality. This is the best we can do in science.

How did we arrive at the present state of natural science, and the case for objective reality? A turning point in my thinking about this grew out of recognition of a commonplace experience in physics: when an interesting idea turns up the odds are fairly good that someone else has already remarked on it, or will, independently, if word does not travel fast enough. An example from another branch of science that even I have known for a long time is the independent recognition of the concept of evolution by natural selection, by Charles Darwin and Alfred Wallace. I have observed other examples in physical science since I was a graduate student, and until recently gave no thought to them. I don't imagine my experience is special in this regard, and indeed I have never heard a physicist remark that the occurrence of multiple discoveries is a phenomenon that could teach us something. But it dawned on me that the fact that Darwin and Wallace both hit on the idea of natural selection reenforces the case that this idea is sensible. After all, two people with feet on the ground independently noticed the evidence. In a similar way, the remarkably large number of multiple discoveries in explorations of the large-scale nature and evolution of the universe suggests that the evidence was offering sensible motivation for the directions in which our thinking was taking us.

My search for who recognized these multiples as a phenomenon to be considered in an assessment of natural science led me to sociologists. They recognize the phenomenon, and I saw that I had something to learn about the practice of physical science from books on sociology, and from that to philosophy, books I had never before thought to consult.

One result was my belated recognition that the sociologists' concept of constructions, social and empirical, accounts for a curious situation in a book I treasure, *The Classical Theory of Fields*, by

Landau and Lifschitz (1951; my copy is the English translation of the 1948 second Russian edition). The first two-thirds of the book is a careful analysis of the basics of the classical theory of electricity and magnetism. This is a well tested and broadly useful theory. The last third of the book presents Einstein's general theory of relativity. I knew from the time I joined Bob Dicke's Gravity Research Group as a graduate student in 1958 that this theory had scant empirical support, quite unlike electromagnetism. So why was general relativity given near equal time to electromagnetism in *The Classical Theory of Fields*? The sociology of constructions reviewed in Section 2.2 led me to understand the situation. It is one of the themes of this book.

I am still relatively unschooled in sociology and philosophy, but see now that there are in these disciplines quite relevant things to say about the practice of science. For those of us who are not sociologists or philosophers, the lessons are best appreciated by application to specific experience of research in physical science. I think of it as a worked example, similar to the problems we set for students with given steps to the solution. Physical cosmology, the study of the large-scale nature of the universe around us, is a good choice for a worked example of research in physical science. The subject in its modern form began with a reasonably clear set of starting ideas a century ago; the pursuit of these ideas involved interesting episodes of confusion and discovery which led to a satisfying conclusion. The tests of this theory are abundant enough, and well enough checked for reliability, to make what most physical scientists who have given it thought agree is a persuasively established case that we have a good approximation to what actually happened as our universe expanded and cooled.

I have to caution that this standard and accepted theory cannot be trusted in an extrapolation forward in time to the remote future, or back in time to arbitrarily large densities. There are interesting ideas about what is going to happen—you will hear talk of the big crunch, the big freeze, and the big rip—and ideas about what was happening before the big bang, whatever that means. But such speculations are beyond the scope of this study of lessons to be drawn from how an example of physical science got to be generally accepted. For the purpose of this book let us be content with the theory of what left the fossils we have been able to identify and interpret.

All natural science is fascinating for those of us with a taste for it. A lot more is added to the case for objective reality from the broad range of well-tested predictions of quantum physics. I recommend the discussion in Steven Weinberg's (1992) book, *Dreams of a Final Theory: The Search for the Fundamental Laws of Nature*. But what is learned from relativity physics and the establishment of the relativistic theory of the evolving universe serves as well to illustrate the argument that we are finding operationally useful approximations to the abstract idea of objective reality. The history and physics of cosmology are simpler, but I see no other reason to favor this example over topics in quantum physics.

The properties of products of living matter play a part in the present story, as examples of the issues that attend the study of complex physical systems. The usual presumption is that living systems are manifestations of the operation of the same objective reality that the evidence from tests of simpler systems indicates is usefully approximated by relativity and quantum physics. But we must live with the fact that a clear demonstration that this is so is beyond present methods of analysis.

There cannot be a universal guide to the nature of science, because operating conditions determine what can be done, and conditions are vastly different in different branches of science. But we can draw useful lessons from a specific worked example, here physical cosmology. It helps that this is a relatively simple branch of natural science. It helps too that I know the subject well; I have been working on it and seen it grow for over half a century.

The story of the development of physical cosmology, presented here in Chapters 3 to 7, is meant to illustrate three things. First, it is an example how physical science is done. Second, it is a guide to thinking drawn from what sociologists and philosophers have been saying we are doing that is particularly relevant from scientists' points of view. Other ideas from these disciplines do not seem so directly relevant, of course. Third, it shows the nature of the argument for objective physical reality that is drawn from the practice of natural science.

I start this story with considerations of the history of thoughts about the nature of the physical sciences tracing back over the past century, to the time when Einstein was musing about the possible nature of a satisfactory theory of gravity and a logically constructed universe. More is to be learned from the still earlier history of science, but coherence is aided by confining attention to later, more directly relevant, developments. Even this limited span of history is rich; I can only offer samples that illustrate ideas that I believe have influenced the way we now think of physical theories within the cultures of science and society. I venture to draw from these examples, and my own experience, a succinct statement of the working assumptions of physics: the ideas that are the fundamental basis for what we do in this subject.

These working assumptions are introduced in Chapter 1 and listed and discussed at the end of the chapter. They are centered on the idea, which is an assumption, that our better theories are useful approximations to objective reality. To some the notion of objective reality is obvious, to others doubtful. My examples in support of the first opinion are drawn from developments in the relativistic theory of the large-scale nature of the universe. As I said, there are many more supporting examples from quantum physics, but the examples from relativity serve.

The social nature of physical science is an important part of scientists' culture. This is recognized more clearly by historians, philosophers, and sociologists than by physical scientists, which should be remedied. I offer the example of how the sociology of physical science and our implicit working assumptions for research in this subject played out in the development and reception of Einstein's general theory of relativity, its eventual establishment by demanding tests, and the development of ideas about its application to the theory of the large-scale nature of the universe. For this purpose, research tends to be most informative in the opening moves. Developments after a particular line of research has been generally adopted as promising tend to be more direct and less instructive, unless the approach fails. I only briefly summarize closing moves on ideas that have proved to be successful so far.

Explanations of nomenclature are needed. I sometimes write about physics, physical science, natural science, or science. I mean by the first the exploration of phenomena that are simple enough to serve as tests of physical theories that are meant to be useful approximations to what we assume is some kind of objective and rationally operating reality. A physical science such as chemistry adds to physics the regularities and theories of phenomenology that are too complex for ready deduction from the approximations to objective reality that physics is claiming to have found. I assume the phenomena of chemistry rest on the same reality as physics, but testing that is challenging. The natural sciences go on from chemistry though geology to botany and biology and on to the nature of living things and the functioning of the human mind. This is an ordering by increasing degree of complexity. I emphasize that it is not meant to be an ordering by merit or interest, in either direction; all are worthy and fascinating examinations of the world around us. As I said, I take cosmology for a worked example of how natural science operates because it is relatively simple and the interaction of theory and observation is fairly easy to judge.

The subject of study in the natural sciences often is called "nature," or "Nature." I use "reality" so I can write "the nature of reality." The few exceptions I take to this rule seem to be demanded by the context. I often use "objective reality," which is redundant but adds emphasis. We have ready ideas about Nature, or reality; it's what we see it around us, it's in your face. I take it that the chair I am sitting on is real, since it seems silly to think I am dreaming. Natural science implicitly operates on the same assumption, that the nature of the world is independent of our ideas about it. Society influences our thinking, but we assume consultation of the empirical evidence corrects misleading thoughts.

I avoid the word "belief" because of the religious connotation, and the words "proof," "demonstration," and "verification," because they seem overly confident for natural science. I use the word "fact" because it appears in interesting discussions to be reviewed, but I mean it with the understanding that in this subject a fact is an approximation, only as good as the supporting data. The same applies to "true," though the word appears uncomfortably often in this text because it is so convenient. I favor "indication" for a reasonable interpretation of an observation, and the awkward but I think accurate terms, "persuasively established" and "a persuasive case," to apply to theories that pass enough tests of predictions to persuade the community to accept the theory, always pending the recognition of new evidence that might force the search for an even better theory. The more abundant the successful predictions, the more persuasive the case for establishment, of course. I use the phrase "community assessment" for ideas that the research community by and large has decided to accept as reasonable, and "standard and accepted" for community ideas that are so well fixed in our minds as to be canonical. The word "canonical" connotes a permanently established case, but of course in science this is not necessarily so.

A "theory" usually is taken to be more completely prescribed and tested than a "model," which tends to be more schematic and speculative. But ideas in cosmology have tended to be intermediate situations. I follow usual practice by taking the words "theory" and "model" to be interchangeable, unless the context suggests particularly schematic ideas that are best characterized as "models." I use the term "physical cosmology" for the theory and observation of the large-scale nature of the universe, the modifier "physical" applied because there are many other kinds of cosmologies.

Space is filled with a near uniform distribution of microwave radiation that is convincingly established to be a remnant, a fossil, from the early universe. Its standard name is the cosmic microwave background, or CMB. I am uneasy about this because the wavelengths of this radiation were much shorter in the past, and while the radiation is a background for us on Earth, coming to us from the sky, it is more properly termed a sea that near uniformly fills space. I call it the "sea of fossil radiation."

The spectra of galaxies are shifted toward the red by the motions of the galaxies away from us. This motion used to be known as the general recession of the galaxies. An observer in another galaxy would see the same general recession; there is no special position in space in the standard cosmology. On average, apart from the motions of galaxies relative to each other produced by local variations of the pull of gravity, the distance between two galaxies, measured by standard rulers and clocks, is increasing at a rate that is proportional to the distance between the two. This is Hubble's law, and the constant of proportionality is Hubble's constant, $H_0$. (That is, the velocity $v$ of recession of a galaxy at distance $r$ is $v = H_0 r$.) The International Astronomical Union has determined that this relation shall be known as the Hubble-Lemaître law, because

Hermann Weyl anticipated it, and Georges Lemaître found a direct prediction, before Edwin Hubble recognized astronomical evidence for the relation. But for brevity I use the old name.

It is sometimes said that space is expanding, but this can be confusing. You and I are not expanding, beyond the usual biological effects, and the galaxies are not expanding, apart from effects of gain of mass by accretion and loss by winds. Best leave it that the spaces between galaxies are increasing, that is, on average the galaxies are moving apart.

Newtonian gravity physics is the nonrelativistic limit of Einstein's general theory of relativity. I use Einstein's term for his theory where it seems appropriate, but more often fall back on the shorter version, general relativity.

The familiar name for the relativistic theory of the expanding universe is the big bang. Simon Mitton (2005) attributes the name, big bang, to Fred Hoyle. The name is not appropriate, because a "bang" connotes an event at a particular place and time. The well-tested cosmology describes a near homogeneous universe, one with no observable edge and no particular place or preferred center, quite unlike a "bang." And cosmology does not deal with an event: it describes the evolution of the universe to its present state from an exceedingly dense, and we now say hot, rapidly expanding early condition. But the name "big bang" is so fixed in people's minds that I use it.

I conclude explanations in this punctilious mode with the difference between precise and accurate measurements. It is relevant for the discussion of the tests of the present standard cosmology, including those reviewed in Section 6.10. Suppose I measure the length of an object many times and find that the average value of the measurements is $L = 1.11 \pm 0.01$ cm. The uncertainty in each measurement is reduced by averaging over many trials. In this example the uncertainty that remains after averaging is give or take about 0.01 cm. But measurements inevitably are affected by systematic errors. Maybe the measuring instrument was slightly bent when I dropped it. Maybe its calibration is a little off. Let us say that my best judgment of the effects of all the systematic errors brings the result to $L = 1.2 \pm 0.1$ cm. The first result, $L = 1.11 \pm 0.01$ cm, is more precise. The second, $L = 1.2 \pm 0.1$ cm, is more accurate. I have been disconcerted to encounter accomplished theorists who fail to grasp the difference. The difference is important.

It is important too to be as accurate as possible about what people have been thinking and doing in this subject. I offer many quotations from authors; better their words than mine. I have been schooled since graduate school days to respect data, and to me quotations are data. Some quotes may be misleading, and other forms of data may be misleading too, but on average data are informative and to be treated with respect. That makes me uneasy about presentations of translations of quotations from German or French. I use published translations when available, and otherwise my translations aided by Google to assist my rudimentary memory of these languages. Something is lost in translation, but I suppose we must live with it.

I include references to the academic literature for those who might like to look into the evidence that supports my claims, or maybe just to be reassured that there is evidence. References to *The Collected Papers of Albert Einstein* (Stachel, Klein, Schulman et al. 1987) in footnotes give the volume and document numbers, available at https://einsteinpapers.press.princeton.edu

Cheryl Misak's (2016) *Cambridge Pragmatism: From Peirce and James to Ramsey and Wittgenstein,* and Karl Sigmund's (2017) *Exact Thinking in Demented Times: The Vienna Circle and the Epic Quest for the Foundations of Science,* introduced me to the varieties of thinking about physics a century ago. Bruno Latour and Steve Woolgar's (1979, 1986) book, *Laboratory Life,* introduced me to the sociologists' arguments for the social aspects of research in natural science. Ernst Mach's *The Science of Mechanics,* in the 1902 and later translations to English, has wonderfully informative discussions of what we now term classical mechanics. They still are well worth reading. Mach's critiques of what he termed the "disproportionate formal development of physics" present us with a fascinating puzzle: what was Mach thinking? I have learned from many other authorities as well, including those listed in the bibliography.

I am on unfamiliar ground in writing about the philosophy of science, always dangerous, and I am particularly grateful to Paul Hoyningen-Huene, David Kaiser, Krystyna Koczanski, Peter

Koczanski, and Cheryl Misak for their reality checks of my thinking about it, though I certainly own the errors that remain. Peter Saulson is a physicist and member of the LIGO Scientific Collaboration that detected gravitational waves from merging astronomical objects. Peter observed a sociologist, Harry Collins, who was observing the physicists in this collaboration. I value Peter's advice from that experience and his thinking about sociology. The sociology of science is a subtle business, and I am grateful to Angela Creager, Regina Kennan, Janet Vertesi, and Harriet Zuckerman for enlightening discussions of their experiences. Michel Janssen, Jürgen Renn, and Cliff Will gave me authoritative advice about the origins of Einstein's general theory of relativity. I am grateful to Charles Robert O'Dell for recollections of research into the astronomy of helium abundances at a critical time for cosmology; Virginia Trimble for recollections of the measurements of gravitational redshifts of white dwarf stars that were important for early tests of general relativity; and Stanislas Leibler for discussion of the curious role of biophysical molecules in this story. I am grateful to Florian Beutler for making for me the critical bottom panel in the figure on page 173. Ingrid Gnerlich, Publisher for the Sciences at Princeton University Press, has given me productive advice on how to make this book more presentable, and she is the source of two reviews by competent readers. Their comments have improved the book, and in particular made me see the need for a firmer explanation of what I was trying to do. Alison Peebles and Lesley Peebles were valuable guides to a preface that explains the purpose of this book. One of my pleasures in writing it has been the considerations of thinking by these people and many others during my long career. In many different ways they were laying out ideas for me to follow.

I had the great good fortune to have landed at a university, Princeton, where I was able to teach physics to interested students and work with inspiring colleagues. Most important to me, and to quite a few others, was Professor of Physics Robert Henry Dicke, Bob to everyone who knew him. I arrived in Princeton from the University of Manitoba as a fresh graduate student in physics in 1958, aiming for research in particle physics. To my great good fortune a fellow Manitoban, Bob Moore, invited me to join him in attending Dicke's research group meetings on how to improve the empirical basis for gravity physics and relativity. Quite a few Nobel Prizes, including mine, grew out of Bob's group. They are a testimony to Bob's vision.

The central administration at Princeton University has not expressed concern about my spending so much time and effort on research that is not at all likely to be monetizable. I count it as an illustration of the value society places on curiosity-driven explorations of what is going on in the world around us, in places large and small.

An active scientist might consider spending a few evenings with this book, reading about what I think is at the core of what they have been doing, and comparing it to their own opinions, which I expect they will have formed even if they never paused to think about it. People who are interested in what is happening in physical science but are not schooled in the mathematical language of the subject may well have spent more time wondering about what we are doing than those of us in the trenches. I have aimed to make the discussion accessible and informative for this camp. There are very few equations in the text, a few more are sequestered in parentheses and footnotes, and I use footnotes to explain the minimal technical jargon that is difficult to avoid. Professional philosophers and sociologists of science will be familiar with the lessons I draw from the wealth of ideas in their subjects, but I hope they will be interested in seeing how I apply my selections to the experience of those of us with feet on the ground.

# CHAPTER ONE

# On Science and Reality

The physical sciences that have grown out of curiosity-driven research have given us the enormous range of technology that so broadly affects our everyday lives, from electric power to the dubious benefit of cell phones. But are the theories that inform all this technology to be considered handy summaries, ways to remember useful experimental results? Or might we accept the assumption that most of us working on research in natural science take for granted, that our well-tested theories are good approximations to a reality that is objective, independent of our attempts to look into what this reality might be?

What is the meaning of reality? It is easy to say that it is what I experience when I wake up. But maybe I am still dreaming, or maybe, as the philosopher Gilbert Harman (1973, page 5) put it, "a playful brain surgeon might be giving [me] these experiences by stimulating [my] cortex in a special way." The thought is playful but the lesson is serious: natural scientists cannot prove they are discovering the nature of objective reality. The argument I am presenting in this book is that physical scientists are in a position to make a persuasive case about what they feel they have learned about a postulate: reality.

You will not often hear the case for reality made by scientists; they would rather go on with research conducted by the precepts they learned from what others are doing and by what they find works for them. This inattention has aided misunderstandings. Scientists point to the demonstrated power of theories that bring to order large ranges of phenomena, and successfully predict a lot more, down to cell phones. But philosophers and sociologists can point out that the best of our scientific theories are incomplete and rest on evidence that is limited by inevitable measurement uncertainties. How then can scientists claim to be discovering absolute truths? When scientists make such claims they should not, cannot, really mean it. The argument instead is that the predictive power of science, demonstrated by all the technology surrounding us, is what one would expect if objective physical reality operated by rules, and if we were discovering useful approximations to these rules.

We should pause for a little closer consideration of the thinking about the predictive power of theories. Suppose a theory is devised to fit a given set of observations. If the theory is a good approximation to what we are assuming is the reality underlying the observations, then we expect that applications of the theory to other, different, situations successfully predicts results of observations of the new situations. The greater the variety of successful predictions, that is, the greater the predictive power, the better the case that the theory is a good approximation to reality. It cannot be a proof; scientists will never be able to claim that the predictive power of their theories demonstrates that they are exact representations of reality. We can only assert that the impressive success of physical science, the broad predictive power of our theories, makes a case that our science is a good approximation to reality that is difficult to ignore.

You may say that these successful predictions are easy to ignore; just do it. But if you do, I urge you to pause to consider the technology you see operating around you. Scientists and engineers can make electrons do their bidding in your cell phone, by the operation of electric and magnetic fields that manipulate electrons and solid and liquid crystals. Does this look like the application of a myth peculiar to our culture? I put it that the many examples of technology of this sort that you see in operation around you make a case for culture-free physics that is very hard to ignore.

If it is accepted that the results of natural science are useful approximations to objective facts, then are our physical theories simply ways to remember these facts? The position taken in natural science is that the predictive power of the well-established theories demonstrates that they are more than that; they are useful approximations to the way reality operates. This is best explained by an example. The one to be considered in the following chapters is physical cosmology, the study of the large-scale nature of the observable universe.

Thinking about the power and limitations of natural science, about empirical facts and their unification and theoretical predictions, is not new. A century ago the American philosopher and scientist Charles Sanders Peirce emphasized the impressive predictive power of the physical theories of the time, what we now term the classical theories of electromagnetism, mechanics, and gravity. But there were expressions of doubt. Another excellent physicist, the Austrian Ernst Mach, asked whether these theories of mechanics and electromagnetism, and of heat and light, are overelaborate, maybe a little artificial. He preferred to think of theories as means of remembering facts. At that time the German-British philosopher F. C. S. Schiller went further, asking whether these facts are only constructions peculiar to eventualities of choices made by our particular society. The predictive power of our theories is even greater now but some things have not changed; we still hear such doubts. This is at least in part because scientists do not usually acknowledge that they work with some constructions that owe more to society than empirical evidence, to say nothing about even less well grounded speculations that physicists occasionally take more seriously than calm consideration would recommend. I will be discussing examples, and will argue that we find common ground by considering the many results from applications of theory and practice in natural science that certainly look like good approximations to reality, while bearing in mind lessons scientists can draw from what sociologists and philosophers observe scientists doing.

My worked example of all this is the growth of physical cosmology, the study of the nature of the universe on the largest scales we can observe. I start with the situation a century ago, when Einstein was thinking about this and he and others were contemplating the broader question of the nature of physical science: how is it done, and what do the results mean? My account begins in Chapter 3 with considerations of Einstein's theory of gravity, general relativity. The evidence we have now is that this theory gives a good description of the expansion of our universe. The idea was discussed in the 1930s, but until the 1960s the evidence for the expanding universe was meagre, the idea of cosmic expansion largely speculative. We can term it a social construction, to take a term from sociology. This with other assessments of science from the perspective of sociology is the subject of Chapter 2. The rest of the present chapter reviews thoughts about the nature of science, now and a century ago. These first two chapters are meant to introduce the considerations that are illustrated by the examples from general relativity and cosmology presented in Chapter 3 and on.

## 1.1   Thinking a Century Ago

A century ago Albert Einstein was thinking about about the nature of a satisfactory theory of gravity, and that led him to wonder how a philosophically sensible universe might be arranged. What were others thinking about physical science then? We see one line of thought in an essay published in *The Popular Science Monthly*,[1] by the American philosopher and scientist Charles Sanders Peirce[2] (1878a pages 299–300). Peirce wrote that[3]

> all the followers of science are fully persuaded that the processes of investigation, if only pushed far enough, will give one certain solution to every question to which they apply it. One man may investigate the velocity of light by studying the transits of Venus and the aberration of the stars; another by the oppositions of Mars and the eclipses of Jupiter's satellites; a third by the method of Fizeau; a fourth by that

1. The magazine continued publication, at least until recently, under the title *Popular Science*.

2. Brent (1993) describes Peirce's creativity and intelligence and not so edifying private life. Peirce's thinking is impressively sensible from my conditioned point of view, but of course there are exceptions. For example, I argue on page 12 that Peirce's discussion of free will is remarkably perceptive, but I do not know what to make of Peirce's thoughts that follow on how to "insert mind into our scheme" that Brent describes on page 208 of his book. But Peirce's ideas are of far more lasting value than most.

3. Because I admire so many aspects of Peirce's thinking, I can't let Peirce's use of gendered pronouns in this quotation and others to follow pass without comment. If challenged to consider the evidence, would Peirce agree that non-gendered pronouns are appropriate in this context? His creativity leads me to hope he could be persuaded to accept this thought that many of us have come to only fairly recently.

of Foucault; a fifth by the motions of the curves of Lissajoux; a sixth, a seventh, an eighth, and a ninth, may follow the different methods of comparing the measures of statical and dynamical electricity. They may at first obtain different results, but, as each perfects his method and his processes, the results will move steadily together toward a destined centre. So with all scientific research. Different minds may set out with the most antagonistic views, but the progress of investigation carries them by a force outside of themselves to one and the same conclusion. This activity of thought by which we are carried, not where we wish, but to a foreordained goal, is like the operation of destiny. No modification of the point of view taken, no selection of other facts for study, no natural bent of mind even, can enable a man to escape the predestinate opinion. This great law is embodied in the conception of truth and reality. The opinion which is fated[1] to be ultimately agreed to by all who investigate, is what we mean by the truth, and the object represented in this opinion is the real. That is the way I would explain reality.

The footnote in Peirce's second last sentence explains that

Fate means merely that which is sure to come true, and can nohow be avoided. It is a superstition to suppose that a certain sort of events are ever fated, and it is another to suppose that the word fate can never be freed from its superstitious taint. We are all fated to die.

We see Peirce's pragmatic endorsement of the idea of objective facts. It is in line with common experience: hit a wine glass and it will break. The idea of facts is taken seriously if usually implicitly in the normal practice of natural science. Also usually implicit in natural science, though Peirce makes it explicit, is the assertion or, better, the postulate that there is "truth and reality" that would be "ultimately agreed to by all who investigate."

In other versions of this essay, which I think appeared later, the first sentence reads "all the followers of science are animated by a cheerful hope ...," and the third from the last begins "This great hope is embodied ..."[4] Misak (2013, pages 50–52) makes it

4. This version is in the Collected Papers of Charles Sanders Peirce, edited by Hartshorne and Weiss (1934), Vol. 5, Book 2, Paper 5, §4, par. 407.

clear that Peirce took the word "hope" seriously. An example is his comment that[5]

> the only assumption upon which he can act rationally is the hope of success . . . it is always a hypothesis uncontradicted by facts and justified by its indispensibleness for making any action rational.

This is a good way to characterize the pragmatic study of physical science and the search for reality. We assume reality operates by rules so we can hope to discover them. It has worked well so far.

Peirce points out that values of the speed of light derived from quite different methods of observation, and using different theories for the reduction of the data from the observations, agree within reasonable allowance for measurement uncertainties. That is, given the result of one method, you could successfully predict what the results from the other methods would be. This demonstration of successful predictions is at the core of the meaning of results of research in physical science, so important that we should take the time to review the experiments and observations Peirce had in mind.

A transit of Venus is observed as a small black dot that moves across the sun along a chord of the face of the sun. Observers at different latitudes on Earth see the transit at chords of different length, meaning they find different times of transit as Venus enters and then leaves the face of the sun. Surveyors had measurements of the radius of the earth, so the distances between observers at different latitudes were known. Newton's theory of the motions of the planets gave the ratio of distances to Venus and the sun. With these data, trigonometry gives the earth-sun distance and the speed of the earth around the sun.[6] The former is known in the jargon as the solar parallax: the angular size of the earth at the sun. The latter is checked by the time it takes for the earth to complete one

---

5. These remarks are from *Writings of Charles S. Peirce: A Chronological Edition*, Volume 2: 1867-1871, page 272. Bloomington: Indiana University Press

6. In more detail, surveyors could establish the physical distance $d$ between two observers normal to the line of sight to the sun. Newtonian mechanics gave the ratio $r_v/r_s$ of the distances to Venus and to the sun. Then the physical distance between the two chords normal to the line of sight is $d(r_s - r_v)/r_v$. That with a little geometry gives the radius of the sun. The observed angular size of the sun gives $r_s$.

orbit around the sun, one year, given the solar parallax. Finally, the time of passage of Venus across the face of the sun gives the speed of Earth relative to the sun. The speed of Earth relative to the speed of light gives the angle, or aberration, by which the apparent positions of stars move as the earth swings around the sun. (Motion at speed $v$ perpendicular to the direction to the star causes its angular position to shift in the direction of motion by $v/c$ radians, where $c$ is the speed of light.) Since our speed around the sun was measured, this ratio gave a measure of the speed of light.

Peirce mentioned a second measure of the solar parallax, derived from observations when Mars, Earth, and the sun are nearly in a line and Mars is on the opposite side of Earth from the sun. Mars is said to be in opposition to the sun. Since Mars is closest to us then, this is a good time to measure the distance to Mars by measuring angular positions of Mars relative to distant stars from different places on Earth, or from one place in the morning and evening when Mars and stars close to it in the sky are just visible. Then again trigonometry with the radius of the earth translates these angles to the Mars-Earth distance. Newtonian physics gives ratios of distances in the solar system, so the Mars-Earth distance gives another measure of the solar parallax, and from that aberration gives the speed of light. The consistency of measurements of the speed of light from observations of transits of Venus and oppositions of Mars checks that Newtonian gravity physics has the ratios of distances of the planets about right, and it checks the always present chance of systematic errors in these subtle measurements.

Peirce also mentioned observations of the eclipses of Jupiter's satellites as they pass behind Jupiter. The timing of the observed eclipses of the moons depends on the Jupiter-Earth distance, because of the light travel time. Given the solar parallax and Newtonian physics to get the Jupiter-Earth distance, the difference of timing of the moon passages as Jupiter moves closer and further from Earth yields the speed of light.

The laboratory measurements by Fizeau, Foucault, and Lissajoux also found the time that light takes to travel a known distance, but in experiments on Earth. The idea is the same as for the timing of the orbits of Jupiter's satellites, but the distances are so very different that I count this as an independent approach.

Peirce mentioned "statical and dynamical electricity." The theories of the electric field of a static distribution of electric charge, and the magnetic field of a steady current of electric charge, are similar apart from a multiplicative factor. That factor was known to be consistent with the speed of light. James Clerk Maxwell found that his theory predicts that electromagnetic waves propagate at this speed of light. As Peirce was writing this essay, experiments by the physicist Heinrich Rudolf Hertz were confirming the existence of these waves. Peirce was referring to some or all of this.[7]

Peirce later added that[8]

> All astronomers, for example, will agree that the earth is ninety-two or ninety-three millions of miles from the sun. And yet, one would base his conclusion on observations of the passage of Venus across the sun's disk; another upon observations of the planet Mars; another upon experiments upon light combined with observations of the satellites of Jupiter.

We see here and in the earlier quotations from Peirce two points that are of prime importance to understanding the nature of research in physical science.

First, Peirce stated that repeated measurements made under repeatable conditions, though they may be by different people, produce the same result. This seems obvious, it is our common experience, but we have not been issued a guarantee, so the evidence of repeatability from experience is essential. There is the complication that experimental and observational scientists, and scientists like me who live by their results, are conditioned to worry about systematic errors in subtle measurements. We pay close attention to what happens, in Peirce's terms, "as each perfects his method and his processes." Peirce was confident that "the results are found to move steadily together." Maybe he had in mind the reduction of corruption by suppression of systematic errors that always are present but in some cases can be made really small. Maybe he also had in

---

7. The electrostatic unit defined by the expression for the force between two elements of charge, and the magnetostatic unit by the expression for the force between two elements of current. The ratio $c$ of the two has units of velocity. This was an experimental result before the connection with the speed of propagation of electromagnetic waves was recognized.

8. Hartshorne and Weiss 1934, Vol. 7, Book 2, Ch. 5, §3, par. 333.

mind that the result is independent of who made the measurement, within the uncertainty. Both are what would be expected under the assumption that reality operates in a lawful and regular way.

The second point Peirce was making is that measurements obtained by different kinds of observations, and reduced by application of different theories, can produce consistent results. By my count Peirce mentioned four independent ways to measure the speed of light: (1) transits of Venus and oppositions of Mars; (2) time delays in orbits of Jupiter's moons; (3) laboratory measurements of light travel times, the same effect as for Jupiter's moons but on length scales so very different as to count as an independent situation; and (4) laboratory experiments with electric and magnetic fields. If we had only one of these measurements then we could say that the speed of light need be nothing more than an artifice, designed to make a story that fits the evidence. But the speed of light determined by one of these methods successfully predicts the results of each of the other three. Peirce also pointed out that the solar parallax derived from one of these methods successfully predicts the solar parallax obtained by a second one. This is an impressive variety of successful predictions, based on applications of different physical theories to different kinds of observations. The consistency of results, within measurement uncertainties, supports the case that the speed of light and the solar parallax are meaningful physical concepts. This is what would be expected if our physical theories were useful working approximations to the behavior of an objective reality that operates by laws we are discovering.

Peirce put it that[9]

> Such is the method of science. Its fundamental hypothesis, restated in more familiar language, is this: There are real things, whose characters are entirely independent of our opinions about them. ... Experience of the method has not led me to doubt it, but, on the contrary, scientific investigation has had the most wonderful triumphs in the way of settling opinion.

I feel the same way.

9. These remarks are from *Writings of Charles S. Peirce: A Chronological Edition,* Volume 3: 1872-1878, page 254

Gerald Holton (1988) gave the same argument a century after Peirce, in *Thematic Origins of Scientific Thought*:

> The meaning behind the statement that we "believe in the reality" of electrons is that the concept is at present needed so often and in so many ways—not only to explain cathode rays, the phenomenon that leads to the original formulation of the concept, but also for an understanding of thermionic and photoelectric phenomena, currents in solids and liquids, radioactivity, light emission, chemical bond and so on.

The progress of science allowed this even broader range of successful predictions, and Holton could have chosen many other examples.

Might there be another theory that would equally well fit these observations, perhaps when considered in the context of some other culture? It cannot be disproved, but for the examples Peirce and Holton mention it seems so unlikely as to be uninteresting. Herbert Dingle's (1931) succinct statement is that "Nature appears to be intelligible."

To repeat the important point: the consistency from diverse measurements and observations in the examples Peirce and Holton gave is what one would expect if objective physical reality operated by rules we can discover, and if the theories employed in interpreting these measurements were good enough approximations to these rules for successful predictions of new phenomena. Examples drawn from research in relativity and physical cosmology are discussed beginning in Chapter 3. But let us consider here the thinking of some of Peirce's contemporaries.

Peirce (1907) recalled that

> It was in the earliest seventies that a knot of us young men in Old Cambridge, calling ourselves, half-ironically, half-defiantly, "The Metaphysical Club"—for agnosticism was then riding its high horse, and was frowning superbly on all metaphysics—used to meet . . .

Misak (2013) shows that Peirce was particularly impressed by Chauncey Wright's contributions to discussions at the Metaphysical Club. Misak (page 17) presents an example of Wright's thinking:

But whatever be the origin of the theories of science, whether from a systematic examination of empirical facts by conscious induction, or from the natural biases of the mind, the so-called intuitions of reason, what seems probable without a distinct survey of our experiences— whatever the origin, real or ideal, the *value* of these theories can only be tested ... by deductions from them of consequences which we can confirm by the undoubted testimony of the senses.

These comments about the importance of what Wright termed "verification" in natural science agree with recent thinking; we need only add that the "testimony of the senses" is now received from quite sophisticated detectors. But I have not found evidence that Wright recognized Peirce's deeper point, the significance of the impressive predictive power of the physical theories of electromagnetism and Newtonian mechanics and gravitation.

William James (1907), another member of The Metaphysical Club and Peirce's close associate for many years, offered a different opinion of physical science. James wrote that

as the sciences have developed farther, the notion has gained ground that most, perhaps all, of our laws are only approximations. The laws themselves, moreover, have grown so numerous that there is no counting them; and so many rival formulations are proposed in all the branches of science that investigators have become accustomed to the notion that no theory is absolutely a transcript of reality, but that any one of them may from some point of view be useful. Their great use is to summarize old facts and to lead to new ones. They are only a man-made language, a conceptual shorthand, as some one calls them, in which we write our reports of nature; and languages, as is well known, tolerate much choice of expression and many dialects.... Such is the large loose way in which the pragmatist interprets the word agreement. He treats it altogether practically. He lets it cover any process of conduction from a present idea to a future terminus, provided only it run prosperously. It is only thus that 'scientific' ideas, flying as they do beyond common sense, can be said to agree with their realities. It is, as I have already said, as if reality were made of ether, atoms or electrons, but we mustn't think so literally. The term 'energy' doesn't even pretend to stand for anything 'objective.' It is only a way of measuring the surface of phenomena so as to string their changes on a simple formula. ... Clerk

Maxwell somewhere says it would be "poor scientific taste" to choose the more complicated of two equally well-evidenced conceptions; and you will all agree with him. Truth in science is what gives us the maximum possible sum of satisfactions, taste included, but consistency both with previous truth and with novel fact is always the most imperious claimant.

Maybe the last sentence in these quotations, and earlier in James's comment that the "great use [of theories] is to summarize old facts and to lead to new ones," is a recognition of Peirce's point that tests of predictions can make a persuasive case for a physical theory. The rest of this commentary does not encourage the idea, however, and nor does James's (1907, page 153) remark that

'The true,' to put it very briefly, is only the expedient in the way of our thinking, just as 'the right' is only the expedient in the way of our behaving.

Misak (2013) reviews Peirce's objection to James's use of the term "pragmatism" to characterize this "expedient" reading of physical science. Peirce's "pragmatism" signifies the search for useful approximations to the way things are, as in electromagnetism and Newtonian physics.

In his book, *The Principles of Psychology*, James (1890, page 454) wrote that

the whole feeling of reality, the whole sting and excitement of our voluntary life, depends on our sense that in it things are *really being decided* from one moment to another, and that it is not the dull rattling off of a chain that was forged innumerable ages ago. This appearance, which makes life and history tingle with such a tragic zest, *may* not be an illusion. As we grant to the advocate of the mechanical theory that it may be one, so he must grant to us that it may *not*. And the result is two conceptions of possibility face to face with no facts definitely enough known to stand as arbiter between them.

The metaphor of a chain of instructions is delightful, as is Peirce's (1892) way of putting it:

The proposition in question is that ... given the state of the universe in the original nebula, and given the laws of mechanics, a sufficiently powerful mind could deduce from these data the precise form of every curlicue of every letter I am now writing.

It seems typical that Peirce (1892) was willing to decide between the ideas of free will and the mechanical theory:

> the conclusions of science make no pretense to being more than probable, and considering that a probable inference can at most only suppose something to be most frequently, or otherwise approximately, true, but never that anything is precisely true without exception throughout the universe, we see how far this proposition [the mechanical theory] in truth is from being so postulated.

We must bear in mind that James was more interested in the ways people behave than in basic physics. Peirce was more on the side of objective reality, to be approached in successive approximations, as we see from the comparison of their thinking about free will.

Despite the hazard of over-interpreting what Peirce might have been thinking, I consider Peirce's comment to be a significant step to the recognition that in classical physics a system can lose memory of initial conditions. That surely would include memory of any programmed instruction about what I am supposed to do next. James's "tragic zest" of life remains a deep puzzle, but I avoid further discussion by confining attention to the far simpler issue of what we learn from tests of basic physics in controlled situations, along the lines of Peirce's discussion of measurements of the speed of light.

The philosopher Ferdinand Canning Scott Schiller (1910, page 89) was even less enthusiastic than James about Peirce's position. Schiller wrote that

> The 'independence' claimed for truth loses all meaning when its ambiguities are analysed. If 'independent' means 'wholly unaffected by,' it stands to reason that truth cannot be independent of us. Two strictly independent things could not co-exist in the same universe. Nor again can truth be 'independent' in the sense of 'unrelated'; for how in that case could we know it? Truth is meaningless if it does not imply a twofold relation, to a person *to whom* it is true, and to an object *about which* it is true. Any 'independence' which ignores either relation is impossible; any which is less than this, is not independence at all.

This is logical, and we have a new aspect of the situation. When I observe a system in a pure quantum state I might change it to another state. How does that affect the state vector of the universe, if there is such a thing? I place the question, with the meaning of free will, outside the bounds of this book.

Peirce's (1903) reaction to Schiller's remark was that

> Mr. Schilller does not believe there are any hard facts which remain true independently of what we may think about them. He admits it requires a hard struggle to make *all* facts suit our fancy, but he holds that facts change with every phase of experience, and that there are none which have been "all along" what history decides they shall have been. This doctrine he imagines is what Professor James means by the "will to believe." He is resolved that it shall have been so.

The problem in this exchange of positions seems to be the failure to recognize the two aspects of a physical theory. William James wrote that theories are "only approximations." Peirce agreed; we see it in the remark about free will. Peirce and James were right; our theories were then and are now approximations. But the best of them are excellent approximations with great predictive power. Peirce presented examples of this aspect of physics, examples that are hard to dismiss. We must consider the other aspect, however, that there are situations, then as now, in which our best theories fail or otherwise are incomplete. Maybe Schiller had in mind something like the incompleteness of theories, and maybe we can put a similar gloss on James's remark that it is "as if reality were made of ether, atoms or electrons, but we mustn't think so literally." Or maybe Schiller and James simply failed to grasp the significance of Peirce's point, that our better theories demonstrate great predictive power even though they are not exact.

Wright, Peirce, and James are considered to be founders of the philosophy of pragmatism.[10] Peirce (1878a, page 293) put it that

10. Peirce was one of the founders of pragmatism. He recalled (Hartshorne and Weiss 1934, Vol. 6, Book 2, Ch. 3, §5, Par. 483) using the term in private discussions, and expressing his thinking along this line in his discussion of the independent ways to measure the speed of light, though he used the term in his writings only later. Menand (2001, page 350) credits William James with bringing attention to Peirce's pragmatism, in a lecture at the University of California in Berkeley in 1898.

we should consider

> what effects, which might conceivably have practical bearings, we conceive the object of our conception to have. Then, our conception of these effects is the whole of our conception of the object.

William James (1907, page 45) asked:

> Is the world one or many?—fated or free?—material or spiritual?—here are notions either of which may or may not hold good of the world; and disputes over such notions are unending. The pragmatic method in such cases is to try to interpret each notion by tracing its respective practical consequences. What difference would it practically make to anyone if this notion rather than that notion were true? If no practical difference whatever can be traced, then the alternatives mean practically the same thing, and all dispute is idle. Whenever a dispute is serious, we ought to be able to show some practical difference that must follow from one side or the other's being right.

In his book, *The Metaphysical Club*, Menand (2001, page 351) wrote that

> Pragmatism is an account of the way people think—the way they come up with ideas, form beliefs, and reach decisions. What makes us decide to do one thing when we might do another thing instead? The question seems unanswerable, since life presents us with so many choices.... Thinking is a free-form and boundless activity that leads us to outcomes we feel justified in calling true, or just, or moral.

In her book, *The American Pragmatists*, Misak (2013) put it that

> Pragmatists are empiricists in that they require beliefs to be linked to experience. They want their explanations and ontology down-to-earth (natural as opposed to supernatural) and they require philosophical theories to arise out of our practices.... The nature of the requisite connections between beliefs and experience is a complex matter for the pragmatists ... [among other reasons because] we need to assume things if we are to go on with our practices.

And among these complex matters for pragmatists was the division between Peirce, who expressed the hope that we can find good approximations to objective facts, and James, who does not seem to have been so sure about that.

My summary impression is that the pragmatism philosophy amounts to the pragmatic acceptance of the world as we observe it. By this definition I find that I have been a pragmatist.[11]

A contribution to divisions of thinking about reality was the advances in physics in the second half of the nineteenth century and continuing through the twentieth. Some found them exciting, but others at the time were disconcerted, naturally enough. Maxwell's replacement of the mechanical picture of the ether by the electromagnetic field was a distinct departure from the familiar laws of mechanics. Boltzmann's account of the second law of thermodynamics, the increase of entropy, applied familiar notions of mechanics, but to a hypothetical particle, the atom, in what may have seemed to be an abstract statistical version of mechanics. Thomson's demonstration of the deflections of electrons by electric and magnetic fields introduced evidence of an unexpected particle to replace the electric fluid, and it established the ratio of the electron charge to its mass. Millikan's 1909 experiment was soon to give evidence of the unique value of the charge of the electron. And consider Max Planck's (1900) introduction of quantum physics.

The Peirce side of the spectrum in the pragmatism camp would find all these developments exciting, but perhaps they added to the less than secure feelings about the ether, atoms, and electrons expressed by James on the other side of the pragmatism spectrum. James may have been wondering whether those physicists really had a solid basis for what they were claiming.

11. In the literature of philosophy you will find the argument that Peirce was an anti-realist because, as Hacking (1983) put it, Peirce was content to allow "the real and the true" to be whatever "information and reasoning would finally result in." The Stanford Encyclopedia of Philosophy is a valuable guide to such considerations, freely available if you have access to the internet. The entry on Scientific Realism (Chakravartty 2017) explains that this philosophy recommends acceptance of "both observable and unobservable aspects of the world described by the sciences." Anti-realists would argue for a more cautious, pragmatic, approach to our theories. This subtle distinction is interesting but the term "anti-realism" is likely to confuse a pragmatic scientist who would take a theory that passes an abundance of tests to be reasonably close to realistic. I avoid use of the terms "realist" and "anti-realist" in the philosophers' technical sense.

While physical science has grown far richer in the past century some things remain the same. In the concluding chapter of *The American Pragmatists*, on current pragmatist thinking, Misak (2013) remarks that

> two kinds of pragmatism emerge. One kind tries to retain a place for objectivity and for our aspiration to get things right while the other is not nearly so committed to that.

Peirce, a century ago, fits in the first of these present-day camps; James, the second.

The Austrian physicist and philosopher Ernst Mach occupied a special place in the spectrum of empiricism, because he had a clear and sharp understanding of physical science, and made important contributions to the study of transonic and supersonic flows. But Mach wanted nothing to do with the "disproportionate formal development of physics" (Mach 1902, page 505) in the more speculative theories that Peirce was willing to consider.

Mach's thinking about this is expressed in his book, *The Science of Mechanics*. (I refer to the Mach 1902 revised and enlarged English translation of the German edition, Mach 1883.) This book presents clear and instructive discussions of what we now term classical mechanics. When I taught introductory physics to university undergraduate students I showed them working models that demonstrate physical effects. I had a ready supply accumulated from many years of these demonstrations, but I added several good ones from Mach's descriptions in my 1960 copy of the 6th edition of the English translation. My favorite is the demonstration of the conservation of angular momentum pictured on page 302 in Mach (1902; page 391 in Mach 1960).

Though Mach clearly understood the physics of the time, he had little patience for the abstract side that might be considered to border on the metaphysics he considered idle speculation. To Mach the idea that motion has an absolute meaning is a metaphysical obscurity; velocity and acceleration surely are meaningful only relative to the rest of matter. The thought became known as Mach's principle. It inspired Einstein's Cosmological Principle, the assumption that the universe is the same everywhere, apart from local irregularities such as stars and planets and people. The story of how Mach's arguments led Einstein to this idea, and how the idea eventually was found to pass the test of a considerable network of evidence, is the subject of Chapter 4.

Mach was willing to consider speculative ideas. On page 493 he wrote that

> A thinking being is supposed to live in the surface of a sphere, with no other kind of space to institute comparisons with. His space will appear to him similarly constituted throughout. He might regard it as infinite, and could only be convinced of the contrary by experience. Starting from any two points of a great circle of the sphere and proceeding at right angles thereto on other great circles, he could hardly expect that the circles last mentioned would intersect. So, also, with respect to the space in which we live, only experience can decide whether it is finite, whether parallel lines intersect in it, or the like. The significance of this elucidation can scarcely be overrated. An enlightenment similar to that which Riemann inaugurated in science was produced in the mind of humanity at large, as regards the surface of the earth, by the discoveries of the first circumnavigators.

Mach also was quite aware of the importance of research that might call for revisions of facts. He wrote (on page 79) of an observer who has the

> opportunity to take note of some *new* aspects of the facts before him— of some aspect which former observers had not considered. A rule, reached by the observation of facts, cannot possibly embrace the *entire* fact, in all its infinite wealth, in all its inexhaustible manifoldness; on the contrary, it can furnish only a rough *outline* of the fact, one-sidedly emphasising the feature that is of importance for the given technical (or scientific) aim in view. *What* aspects of a fact are taken notice of, will consequently depend upon circumstances, or even on the caprice of the observer. Hence there always is opportunity for the discovery of new aspects of the fact, which will lead to the establishment of new rules of equal validity with, or superior to, the old.

It is an empiricist's refrain: ideas are only as good as the evidence.

Mach understood the power of theories but saw a restricted role for them. He wrote (on page 481 in Mach 1902) that

It is the object of science to replace, or *save*, experiences, by the reproduction and anticipation of facts in thought. Memory is handier than experience, and often answers the same purpose. This economical office of science, which fills its whole life, is apparent at first glance; and with its full recognition all mysticism in science disappears.

Thus, while Ludwig Boltzmann and others were making good use of the concept of atoms, Mach wrote (on page 492) that

> The atomic theory plays a part in physics similar to that of certain auxiliary concepts in mathematics; it is a mathematical *model* for facilitating the mental reproduction of facts. Although we represent vibrations by the harmonic formula, the phenomena of cooling by exponentials, falls by squares of times, etc., no one will fancy that vibrations *in themselves* have anything to do with the circular functions, or the motion of falling bodies with squares. It has simply been observed that the relations between the quantities investigated were similar to certain relations obtaining between familiar mathematical functions, and these *more familiar* ideas are employed as an easy means of supplementing experience.

Mach was referring to the observation that the distance $d$ a freely moving body falls in time $t$ starting from rest is proportional to the square of the time (that is, $d = gt^2/2$, where $g$ is the local acceleration of gravity). The "circular functions" (sines and cosines, as in $\cos(\omega t)$) describe the displacement of a simple oscillator as a function of time. This still is a widely used model in physics. As Mach wrote, a fall by the square of the time, a simple oscillator, and an atom are useful constructs in theories. The first two are are not meant to be considered real, they are only helpful models. So Mach was asking a good question: why take atoms to be real, as opposed to yet another helpful model?

In another way to put it, Mach protested against

> the disproportionate formal development of physics [that has led] the majority of natural inquirers [to] ascribe to the intellectual implements of physics, to the concepts mass, force, atom, and so forth, whose sole office is to revive economically arranged experiences, a reality beyond and independent of thought.... A person who knew the world only through the theatre, if brought behind the scenes and permitted to view

the mechanism of the stage's action, might possibly believe that the real world also was in need of a machine-room, and that if this were once thoroughly explored, we should know all. Similarly, we, too, should beware lest the *intellectual* machinery, employed in the representation of the world *on the stage of thought*, be regarded as the basis of the real world.

I admire Mach's analogy, but remain puzzled by his refusal to be impressed, or maybe it was his refusal to admit to being impressed, by the predictive power of what he termed the "economically arranged experiences." Mach's economy of science, also known as our suite of physical theories, predicts new phenomena that are found to pass empirical tests. In Mach's time this was clear enough to Peirce, and now we can cite many more examples such as the one Holton (1988) mentioned.

My empiricist tendencies lead me to sympathize with Mach's dislike of "disproportionate formal development of physics," but Mach's thinking was not prescient. Formal developments gave us the wonderfully productive principles of quantum and relativity physics.

Others shared Mach's thinking. On the subject of electricity and magnetism, Mach's (1902 page 494) opinion was that

> Our conceptions of electricity fit in at once with the electrical phenomena, and take almost spontaneously the familiar course, the moment we note that things take place as if attracting and repelling fluids moved on the surface of the conductors. But these mental expedients have nothing whatever to do with the phenomenon *itself*.

Auguste Comte, who was of the generation before Peirce and Mach, argued for "positivisme scientifique," along lines quite similar to Mach's later thinking. Comte asked (in Comte 1896, the later translation to English)

> What scientific use can there be in fantastic notions about fluids and imaginary ethers, which are to account for phenomena of heat, light, electricity and magnetism? . . . These fluids are supposed to be invisible, intangible, even imponderable, and to be inseparable from the substances which they actuate. Their very definition shows them to have no place in real science; for the question of their existence is not a subject

for judgment: it can no more be denied then affirmed: our reason has no grasp of them at all. Those who in our day, believe in caloric, in a luminous ether, or electric fluids, have no right to despise the elementary spirits of Paracelsus, or to refuse to admit angels and genii.

The American philosopher John Dewey (1903) put it that

not every hypothesis can be actually experienced. For example, one employs in physics the hypothesis of electric fluid, but does not expect actually to meet with it.

We have seen that James expressed a similar sentiment. These authors put an important question: is the existence of electric fluids a subject for judgment?

The scientific community has a standard way to answer the question; the case for or against the existence of hypothetical objects such as atoms and electric fluids is made by checking predictions against measurements. Mach understood the physical science involved, but he refused to recognize the idea of tests by predictions. Consider Mach's (1902 page 599) argument that

Even if an hypothesis were fully competent to reproduce a given department of natural phenomena, say, the phenomena of heat, we should, by accepting it, only substitute for the actual relations between the mechanical and thermal processes, the hypothesis. The real fundamental facts are replaced by an equally large number of hypotheses, which is certainly no gain.

As this is stated, the theory of heat—thermodynamics—would be a circular construction: the theory fits what is observed because it is constructed to fit what is observed. But there is more to the situation than this; Maxwell's relations, the predictions of relations among useful thermodynamic quantities, are a real and useful gain. But let us consider an example that is more direct and counter to Mach's thinking on this point.

Mach (1902 pages 157, 231, 302) explained how the Newtonian physics of mechanics and gravity account for the motions of the planets, the motions of the moons around their planets, the flattened shape of the rotating earth, ocean tides and trade winds, the rotation of the plane of Foucault's pendulum, and the operation of a pendulum clock. This is a wonderfully diverse variety of phenomena, operating on distance scales that range from roughly $10^{13}$ cm for the paths of the planets down to a few centimeters for a pendulum. All are fit by the brief statements of Newtonian physics. Mach termed this an example of the economy of science. The more adventurous way to put it is that Newtonian physics seems to be a good approximation to how objective reality operates.

Peirce presented another example of this economy, without naming it, in the multiple ways to find the distance to the sun and the speed of light from astronomy and the study of electricity and magnetism. Mach knew all this physics. In the collection of *Popular Science Lectures* Mach (1898) explained measurements of the speed of light in the laboratory and from the timing of the orbits of the moons of Jupiter. Mach also remarked that the

> current which by the magnetic C. G. S. [centimeter, gram, second] standard represents the unit, would require a flow of some 30,000,000,000 electrostatic units per second through its cross-section. Accordingly, different units must be adopted here. The development of this point, however, lies beyond my present task.

This is the speed of light, in units of centimeters per second, that relates laboratory electrostatic and magnetostatic units. It also determines the speed of propagation of Herz's waves. Mach understood the physics, and in these lectures was making the same point as Peirce, but putting a less enthusiastic gloss on it.

Mach's economy of science was then and remains so commonplace that we might fail to notice that it is a remarkable phenomenon. The world around us is observed to be operating by rules we have discovered in approximations that are predictive and unify broad ranges of phenomena. It is remarkable too that although Mach certainly understood this property of experience he made no attempt to understand what it might signify, whether there might be something to learn from it. This economy, also known as the predictive power of our better theories, allows the judgments that Comte felt are missing. The luminous ether Comte distrusted has been discarded in favor of the electromagnetic field that has proved to be far more predictive. The hypothetical electric fluid has become the electrons and ions of the predictive and well-tested theory of electromagnetism. Of course, it is far easier now to say that Mach's economy of physics would suggest that the physical theories in Mach's days were useful approximations to the way reality operates. Mach certainly would not venture this far; I expect he would have dismissed the thought as uninteresting metaphysics.

Peirce was willing to venture further into speculation, willing to entertain the abstract idea of atoms, though with particular attention to ideas that might be tested. This thinking is summarized in his remark that[12]

> I am a physicist and a chemist, and as such eager to push investigation in the direction of a better acquaintance with the minute anatomy and physiology of matter. What led me into these metaphysical speculations, to which I had not before been inclined, I being up to that time mainly a student of the methods of science, was my asking myself, how are we ever going to find out anything more than we now [know] about molecules and atoms? How shall we lay out a broad plan for any further grand advance?

We still ask ourselves, how shall we plan for the next grand advance?

A century ago, recognition of the considerable range of tests and practical applications of mechanics, gravity theory, electromagnetism, and thermodynamics encouraged some to take a more adventurous line of thought, that physical science might be advanced enough to be formulated in a precise logical way. The sixth of David Hilbert's (1900) famous mathematical problems presented at the 1900 *International Congress of Mathematicians* is the "Mathematical treatment of the axioms of physics." Hilbert's explanation begins

> The investigations on the foundations of geometry suggest the problem: To treat in the same manner, by means of axioms, those physical sciences in which mathematics plays an important part; in the first rank are the theories of probabilities and mechanics.

The goal of the Vienna Circle of people who were inspired by Mach's thinking was even broader. They sought a unified system of mathematics, philosophy, physics, and the social sciences,

---

12. Hartshorne and Weiss 1934, Vol. 7, Book 3, Ch. 3, §7, Par. 506

organized in a strictly logical way on the basis of direct empirical facts. Philipp Frank (1949), who became a member of the Vienna Circle, recalls his early thinking as he was moving toward this vision. In the years around 1907

> I used to associate with a group of students who assembled every Thursday night in one of the old Viennese coffee houses. We stayed until midnight and even later, discussing problems of science and philosophy.... At the turn of the century the decline of mechanistic physics was accompanied by a belief that the scientific method itself had failed to give us the "truth about the universe"

Frank recalls that they were fascinated by the assessment of the evolving situation in natural science by the philosopher of science Abel Rey, in the book *La théorie de la physique chez les physiciens contemporains* (Rey 1907). In Frank's translation, Rey argued that

> Science became nothing but a symbolic pattern, a frame of reference. Moreover, since this frame of reference varied according to the school of thought, it was soon discovered that actually nothing was referred that had not previously been fashioned in such a way that it could be so referred.

How to remedy this circular construction? Frank recalls that

> We agreed with Abel Rey's characterization of Poincaré's contribution as a "new positivism" which was a definite improvement over the positivism of Comte and Mill.

Poincaré (1902, page 127) wrote that[13]

> Experience is the sole source of truth: it alone can teach us something new; it alone can give us certainty.

The emphasis on experience rather than a priori knowledge was central to the thinking of Wright and Peirce, and that of Mach and the Vienna Circle. But Poincaré was cautious about how far experience can take us. He remarked that Mariotte's law (also known as Boyle's law) is quite accurate for some gases, but the gas examined in sufficient detail breaks up into the chaotic motions of particles.

13. I use Halsted's (1913) translation.

Poincaré (1902, page 132) suggested that gravity examined in fine enough detail similarly might depart from the simplicity of Newton's law.

> No doubt, if our means of investigation should become more and more penetrating, we should discover the simple under the complex, then the complex under the simple, then again the simple under the complex, and so on, without our being able to foresee what will be the last term.

Poincaré's vision, which I characterize as successive approximations all the way down, seems as plausible as the talk of a final theory. Poincaré (1905, page 14) also asked,

> Does the harmony the human intelligence thinks it discovers in nature exist outside of this intelligence? No, beyond doubt a reality completely independent of the mind which conceives it, sees or feels it, is an impossibility. A world as exterior as that, even if it existed, would for us be forever inaccessible.

We have seen similar thoughts by Schiller (1910). The logic is impeccable, the meaning still open to debate.

Rudolf Carnap (1963), who became another member of the Vienna Circle, recalls his earlier thinking.

> I imagined the ideal system of physics as consisting of three volumes: The first was to contain the basic physical laws, represented as a formal axiom system; the second to contain the phenomenal-physical dictionary, that is to say, the rules of correspondence between observable qualities and physical magnitudes; the third to contain descriptions of the physical state of the universe

This is Hilbert's style, but I think not Poincaré's.

The manifesto, *Wissenschaftliche Weltauffassung Der Wiener Kreis*, or *The Scientific World View of the Vienna Circle* (Hahn, Neurath, and Carnap 1929), gives us a further impression of their approach. It mentions inspiring thinkers, including Bertrand Russell and Ludwig Wittgenstein on the side of mathematics and philosophy, and Ernst Mach and Albert Einstein on the side of physics. Poincaré and Mach were more directly influential in how members of the Vienna Circle thought about physical science.

The concluding paragraph in the manifesto in the section on the Fundamentals of Physics is particularly relevant for our considerations.

> Through the application of the *axiomatic method* to the problems mentioned, the empirical components of science differ from the merely conventional, the meaningfulness from the definition. There is no room left for a synthetic judgment *a priori*. The fact that knowledge of the world is possible is not based on the fact that human reason imprints its form on the material, but on the fact that the material is ordered in a certain way. Nothing can be known from the start about the type and degree of this order. The world could be organized much more than it is; it could, however, also be much less ordered but still understandable. Only the pressing step-by-step research of empirical science can teach us the lawful nature of the world. The method of *Induction*, the conclusion from yesterday to tomorrow, from here to there, is of course only valid if there is a law. But this method is not based on an *a priori* premise of this law. It may be used wherever it leads to fruitful results, whether sufficient or insufficiently justified; it never grants security. But epistemological reflection demands that an induction conclusion is only given meaning insofar as it can be verified empirically. The scientific view of the world will not reject the success of a research work because it has been achieved with inadequate, logically insufficiently clarified or empirically inadequately justified methods. However, it will always strive for and demand verification through thoroughly firm aids, namely the indirect or direct tracing back to experiences.

In the foreword to this document the authors are named as Hans Hahn, Otto Neurath, and Rudolf Carnap. Sigmund (2017) mentions more contributors, so it is not surprising to see expressions of a variety of thoughts in this paragraph.

I take the paragraph to be a statement of what is termed logical positivism, or logical empiricism. The philosophy is no longer considered interesting by some authorities, but I see things to like about these statements. The Vienna Circle seeks the "lawful nature of the world," as we still do. They recognize that it is an assumption that "there is a law." With Poincaré they emphasize that a law requires "verification through . . . the indirect or direct tracing back to experiences." Perhaps they were more confident than Poincaré that the law can be discovered, rather than approached in successive approximations. With Peirce, and the Vienna Circle, we continue to assume there is a lawfully acting reality so we can hope to discover its laws. In short, this paragraph is a reasonable statement of a good part of current thinking about research in physical science. I see two missing points.

First, Carnap and others in the Vienna Circle proposed a "creative reconstruction" that would reduce the language of natural science to a strictly logical and manifestly meaningful system. Research in natural science cannot operate this way because we are attempting to discover the logic by which our assumed reality operates. Informal trial and error is inefficient, but it has served well in our learning how to frame the language that gave you cell phones, despite the incomplete nature of the language.

The second missing point is the importance of empirical tests of theoretical predictions, tests of extrapolations beyond the range of evidence used to construct a theory. We have seen Perice's examples of the impressive predictive power of electromagnetic and Newtonian physics. Why this predictive power? It is what you would expect if these theories were useful approximations to the way reality operates, useful enough to allow extrapolation to different situations. I have not found arguments in the document for or against this thought; perhaps it would have been considered an example of the metaphysical thinking Mach so disliked and members of the Vienna Circle sought to replace with a logical arrangement of the empirical facts, and only the empirical facts.

An approach similar in spirit to the sixth of Hilbert's mathematical problems, and the thinking taken up by the Vienna Circle, is the search for a unified physical theory, maybe beginning with the unified field theory of gravitation and electromagnetism. Einstein and Hilbert looked for it; Kaluza (1921) and Klein (1926) found it in a space of five dimensions. But it adds nothing to either theory if we accept Klein's prescription for a tiny fixed length that closes the fifth dimension. Some argue that we might be on the threshold of finding the "theory of everything," or as Steven Weinberg (1992) puts it, "Dreams of a Final Theory," maybe along the lines of the Kaluza-Klein approach. Weinberg presents a careful and well-informed case; the final theory might serve as the axioms that Hilbert and others dreamed of a century ago. But, as Poincaré argued, we cannot empirically determine that we have the final theory, as opposed to the best approximation the world economy can afford to establish.

The philosopher Karl Popper was not a member of the Vienna Circle, but there are accounts of their mutual influence. Popper's philosophy is commonly remembered for the emphasis on falsifiability. In his autobiography Popper recalls being fascinated by Einstein's remark that, if specific predictions of relativity theory disagree with the measurements, then the theory is untenable. Popper (1974, page 29) recalls that

> I arrived, by the end of 1919, at the conclusion that the scientific attitude was the critical attitude, which did not look for verifications but for crucial tests; tests which could *refute* the theory tested, though they could never establish it.

Popper (1965, page 287) also considered confirmability:

> The severity of possible tests of a statement or a theory depends (among other factors) on the precision of its assertions and upon its predictive power.... The better a statement can be tested, the better it can be confirmed, i.e. attested by its tests.

An illustration of Popper's (1965, page 117) thinking about what tests might tell us about reality is his comment that

> Testable conjectures or guesses, at any rate, are thus conjectures or guesses about reality; from their uncertain or conjectural character it only follows that our knowledge concerning the reality they describe is uncertain or conjectural. And although only that is certainly real which can be known with certainty, it is a mistake to think that only that is real which is known to be certainly real. We are not omniscient and, no doubt, much is real that is unknown to us all.

And let us take note of Popper's (1945, page 208) opinion of the influence of society on science:

It was one of the greatest achievements of our time when Einstein showed that, in the light of experience, we may question and revise our presuppositions regarding even space and time, ideas which had been held to be necessary presuppositions of all science and to belong to its 'categorial apparatus.' Thus the sceptical attack upon science launched by the sociology of knowledge breaks down in the light of scientific method. The empirical method has proved to be quite capable of taking care of itself.

Let us turn to the considerations of the sociology of scientific knowledge.

## 1.2  On Social, Empirical, and Circular Constructions

Thomas Kuhn's thinking about constructions in natural science is presented in the book *The Structure of Scientific Revolutions*. Kuhn wrote, in the postscript in the second edition (Kuhn 1970, page 206), that

A scientific theory is usually felt to be better than its predecessors not only in the sense that it is a better instrument for discovering and solving puzzles but also because it is somehow a better representation of what nature is really like. One often hears that successive theories grow ever closer to, or approximate more and more closely to, the truth. Apparently generalizations like that refer not to the puzzle-solutions and the concrete predictions derived from a theory but rather to its ontology, to the match, that is, between the entities with which the theory populates nature and what is "really there." Perhaps there is some other way of salvaging the notion of 'truth' for application to whole theories, but this one will not do. There is, I think, no theory-independent way to reconstruct phrases like 'really there'; the notion of a match between the ontology of a theory and its "real" counterpart in nature now seems to me illusive in principle.

Physicists have a regrettable tendency to declare that we are at last approaching the final theory, the discovery of the ultimate nature of reality. But physics cannot do that, because all empirical tests have limited accuracy. This means our world picture, our ontology, must evolve with the advance of science. Kuhn was right; we can only claim to establish persuasive cases for approximations to an assumed reality. In the following chapters I argue for some quite persuasive examples.[14]

Consider also the thought in an extract Galison (2016) drew from Kuhn's correspondence:

> objective observation is, in an important sense, a contradiction in terms. Any particular set of observations ... presupposes a predisposition toward a conceptual scheme of a corresponding sort: the "facts" of science already contain (in a psychological, not a metaphysical, sense) a portion of the theory from which they will ultimately be deduced.

This is a good description of the circular constructions of theories that fit our perceptions of the evidence because they were designed to fit our perceptions. It is a standard and important part of research in the natural sciences; a more charitable way to put it is that we attempt to follow the evidence. The follow-up stage is the search for checks of a theory thus constructed by what follows from the theory that was not part of the perceptions that went into its construction. The tests of predictions have been an obsession among physical scientists for a long time.

General relativity and relativistic cosmology pass the broad range of demanding tests of predictions to be discussed in following chapters. So what would this signify in Kuhn's philosophy? He had a PhD in physics, a doctoral dissertation with van Vleck (Kuhn and van Vleck 1950), but later Kuhn displayed little interest in the experimental side of physics. It is illustrated by what Kuhn wrote (on page 61 in *The Structure of Scientific Revolutions*) about

14. David I. Kaiser, in a private communication, writes that "Kuhn was particularly concerned about apparent mismatches between the sets of entities that various theories described as being part of reality. For example, in Maxwell's day there was no empirical evidence for fundamental, microscopic charge-carriers with fixed electric charge—what would later be identified as electrons [and ions]. Yet calculations in quantum electrodynamics begin from an assumption that electrons exist and are among the fundamental entities responsible for electromagnetic phenomena. Given the very different assumptions about what the world is ultimately made of—and the equally striking fact that theoretical predictions from Maxwell's theory and from quantum electrodynamics are compatible in various limits—some philosophers of science have developed arguments for "structural realism": perhaps natural science reveals relationships that really hold in nature, even if theories make incompatible claims about specific theoretical entities. See the entry on "Structural Realism" in the Stanford Encyclopedia of Philosophy (Ladyman 2020)"

the manipulations of theory undertaken, not because the predictions
in which they result are intrinsically valuable, but because they can be
confronted directly with experiment.

This way to put it is quite different from the emphasis on predictions in the natural science camp. It is of course understood that successful predictions need not be expected of a story invented to fit a given set of observations, apart from the occasional fluke. But successful predictions are required of a theory that is thought to be a useful approximation to an assumed rationally operating reality. There have been exceptions to this rule—consider the community acceptance of general relativity in 1960—but I admit none in this discussion. Successful predictions mean a lot to scientists, and examples are abundant in the established branches of the physical sciences. I have not found evidence of appreciation of this by Kuhn.

Kuhn (1970) introduced the influential concept of paradigm shifts: changes in the accepted world picture based on the normal science that was taken to represent reality. Kuhn's historical examples are real and important. But the effects of paradigms that were introduced in physical science in the past century are better described as additions to a layered growth of the world picture of physical science. For example, Maxwell's equations of electromagnetism are more than a century old, and we have a deeper theory, quantum electrodynamics. The quantum version is a profoundly important paradigm, but much remains the same. We still teach the classical limit of quantum electrodynamics, Maxwell's equations, because they are widely applied in science and they are of broad practical use to society. The classical theory has not been replaced, it has become a useful limiting case of the quantum theory, which is in turn a remnant of the theory of electroweak symmetry. The additions of these last two paradigms enrich our world picture, but Maxwell's classical electrodynamics remains an important part of normal science. If Ohm's law occasionally failed when it shouldn't we would be hearing about it.

As will be described, the science of physical cosmology also grew by the addition of layers of ideas: Einstein's 1917 picture of a homogeneous but static universe; the idea in 1930 that the homogeneous universe is expanding; the idea in 1970 that the universe was hot in the early stages of expansion, and left fossils; the ideas in 1990 that the mass of the universe is largely nonbaryonic dark matter and Einstein's cosmological constant is causing the present rate of expansion of the universe to increase; the precision tests in the years around 2000 that settled many debates; and the increased variety and precision of tests since then that drove the community acceptance of a standard and well-tested theory of the large-scale nature of the universe. You might term this a series of revolutions, or paradigm shifts. But the better way to put it is that the science grew by a series of paradigm additions. We hope for move to come; it would be good to have a more empirically based picture of the nature of the nonbaryonic dark matter.

Might there be an actual paradigm shift, maybe to a universe without dark matter? It looks exceedingly unlikely in light of the tests to be considered in Section 6.10. But we have no proof; time will tell.

Kuhn of course accepted the progress of physical science over the centuries that has grown to fit an ever broader range of phenomena with increasing precision and practical applications. And he was right to argue that, despite this, a physicist who causally mentions that something is real must not mean it, strictly speaking. Physics has theories that pass demanding tests, but however precise the tests, they are approximations, and however successful the theories, they are incomplete too.

I have heard it said that physicists objected to Kuhn's thesis when the book appeared, not for the point about circular constructions, but for Kuhn's argument that there is a place for sociology in what is supposed to be impersonal research that is revealing objective reality. I remember looking into the book then, and do not remember any pronounced reaction, or observing any among colleagues in physics. Physicists usually have other things they would rather do. But I suppose the culture of physical science has evolved, because the mutual influence of science and society seems obvious now. This is the subject of Chapter 2.

Popper and Kuhn agree that physicists cannot demonstrate, as a theorem, the existence of objective reality, but Popper does not hesitate to accept the idea of reality while Kuhn emphasizes the role of society in determining what is considered to be reality. Thoughts about reality and society along lines closer to Kuhn than Popper were expressed by Schiller a century ago, and more recently by the philosopher Bruno Latour and the sociologist Steve Woolgar (1979) in their book, *Laboratory Life: The Social Construction of Scientific Facts*. It is an account of what Latour made of his embedding for two years in a research laboratory for molecular biology, the Salk Institute for Biological Studies. Their assessments are discussed in Chapter 2, on the sociology of science. Directly relevant here are two of their remarks. First,

> The argument between realists and relativists is exacerbated by the absence of an adequate definition of reality. It is possible that the following is sufficient: that which cannot be changed at will is what counts as real.

I imagine most physicists today would not be uneasy about this way to put it. It is operational, which is good. The definition changes with advances in science, which one would not expect of objective reality, but there is nothing wrong with that because we are assuming adjustments of the definition mean we are finding better approximations to the nature of reality.

The second remark to consider is the report that

> In no instance did we observe the independent verification of a statement produced in the laboratory. Instead, we observed the *extension* of some laboratory practices to other arenas of social reality, such as hospitals and industry ... This does not imply that the statement holds true *everywhere.*.... It is impossible to prove that a given statement is verified outside the laboratory since the very existence of the statement depends on the context of the laboratory. We are not arguing that somatostatin [a peptide hormone that regulates the endocrine system] does not exist, nor that it does not work, but that it cannot jump out of the very network of social practice which makes possible its existence.

We can compare this report to Peirce's discussion of the "independent verification" of the speed of light by the consistency of results from quite different means of observation. Peirce's example was possible because the measurements, while delicate, are simple to interpret, and they do not seem likely to have been confused by the complexity of the phenomena. The quantitative agreement of measures of the speed of light based on observations of different phenomena analyzed by applications of different theories makes a good case for reality of the speed of light; it impresses me. Latour and Woolgar do not report similar instances of "independent verification" of the studies of massive molecules outside suitably equipped laboratories. The big difference from Peirce's example is that the molecules of biophysics are large, their structures complicated, and their interactions with their environments deeply complex.

We do have a good case for the objective reality of a far simpler molecule, molecular hydrogen. The use of standard quantum physics to compute the structure of molecular hydrogen and its predicted binding energy and energy levels passes precision tests that make a compelling case that this molecule is real, with a secure place among predictions of the canons of physical science. Producing a similar level of consistency of the theory and observations of biophysical molecules is a far greater challenge. The standard thinking among scientists is that this open challenge must be accepted as one of the incomplete aspects of the subject, to be remedied incrementally as methods of analyses improve, though with no guarantee of a convincing answer to whether biophysical molecules fit the fundamental physical theory we have now or hope to improve.

Peirce and Mach expressed confidence in facts, or at least good approximations to them. Latour and Woolgar (1986, page 175) offered a measure of agreement:

> facts refuse to become sociologised. They seem able to return to their state of being "out there" and thus to pass beyond the grasp of sociological analysis.

I think that in this statement Latour and Woolgar are expressing themselves as willing to accept the notion of the reality of facts that are "out there," to be seen by anyone who cares to look, though I imagine they would prefer to say, to be seen by anyone who is properly equipped to be able to look. The same is true of the transits of Venus, of course.

Another sociologist, Robert Merton (1973, page 271), in a commentary on the politicization of science in the Soviet Union, and in particular on an editorial, "Against the Bourgeois Ideology of Cosmopolitanism," was more direct:

> the criteria of validity of claims to scientific knowledge are not mat-
> ters of national taste and culture. Sooner or later, competing claims to
> validity are settled by universalistic criteria.

Mach might have objected to the description of "facts [that] refuse to become sociologised" as overly metaphysical, but I expect that he, with Peirce on the side of physics, and Latour, Woolgar, and Merton on the side of sociology, would be comfortable with the idea of facts that are out there to be discovered, independent of the social norms of those who choose to look. Other sociologists and philosophers might not be so sure about this; discussion of the issue continues in Chapter 2.

## 1.3  Philosophies of Science

Our agreed-upon laws of logic and mathematics frame physical theories that make predictions that fit observations remarkably well. This statement is commonplace; the practical results have transformed our way of life. It also is a profoundly important fact about the nature of the world around us. What might we make of it? Peirce's (1878b) thought is that[15]

> It seems incontestable, therefore, that the mind of man is strongly
> adapted to the comprehension of the world; at least, so far as this goes,
> that certain conceptions, highly important for such a comprehension,
> naturally arise in his mind; and, without such a tendency, the mind
> could never have had any development at all.... It must be admitted
> that it does not seem sufficient to account for the extraordinary accu-
> racy with which these conceptions apply to the phenomena of Nature,
> and it is probable that there is some secret here which remains to be
> discovered.

An example of Einstein's (1922b, p. 28) thinking in this direction is that

15. Peirce (1878a) is the second paper in the series, *Illustrations of the Logic of Science*. Peirce (1878b) is the fifth paper in this series.

an enigma presents itself which in all ages has agitated inquiring minds. How can it be that mathematics, being after all a product of human thought which is independent of experience, is so admirably appropriate to the objects of reality? Is human reason, then, without experience, merely by taking thought, able to fathom the properties of real things.

Also from Einstein (1936),

One may say "the eternal mystery of the world is its comprehensibility."

A more recent example of this way of thinking is displayed in the title of Eugene Wigner's (1960) essay, *The Unreasonable Effectiveness of Mathematics in the Natural Sciences*, and his comment in the article about

the two miracles of the existence of laws of nature and of the human mind's capacity to divine them.

As Peirce remarked, we do seem to be "adapted to the comprehension of the world." We, like other animals, are born with an intuitive a priori appreciation of elements of mechanics. I offer the example of the reliability of the operation of a mechanical lever, as in knee bends. And I suppose we can take it that this primitive, apparently a priori, knowledge of mechanics is the result of experience built in through the course of Darwin's evolution that produced the present species. Perhaps our a priori appreciation of logic can be attributed to the adaptive value of recognition of hard facts on the ground, learned along with the survival of the fittest. Perhaps our appreciation of mathematics follows from the learned value of logic.

There are apparent counterexamples to a logically operating universe. Why was this home hit by lightning and that one spared? Why did this person die young while that one lived to an old age? Somewhere along the course of history people had to have grasped the great difference between the complexity of such questions and the simplicity of experiences that lend themselves to tests of reproducibility. Reproducibility is an encouraging sign of reality that operates in a logical way we are conditioned to recognize, but going even this far is an assumption, of course.

Many lines of philosophical thinking touch on the "comprehension of the world," real or constructed. For the present discussion of what people have been thinking about science we are allowed considerable simplifications.[16] Let us take Ernst Mach, and the Vienna Circle in the years around 1930, to serve as prototypes in the tradition of logical positivism, or empirical positivism, or simply positivism. This tradition acknowledges the success in bringing phenomena into reasonably good order, at least after a fashion, as is sensible, because we see and use the products of physical science in our everyday lives. Mach certainly understood this success, and termed it the economy of science, or sometimes the economy of thought. But Mach and others in this tradition would not venture into speculation about what this kind of economy might signify. I do not understand this side of Mach's thinking, because his "economy" is a remarkable phenomenon, well established, which has taught us a lot about the operation of the world around us.

Let us take Charles Sanders Peirce to be a prototype for the second tradition, pragmatism, discussed on page 14. As Peirce practiced it, pragmatism accepts the advances that positivist thinking has brought us and adds the exploration of hypothetical notions such as atoms and molecules, or more recently quarks, nonbaryonic dark matter, and even multiverses. William James also was a pragmatist, by the conventional definition, but with a difference. Peirce took seriously what we now term the predictive power of theory; James was not so sure. Natural scientists still tend to operate by Peirce's version of pragmatism, usually without thinking about it, apart from occasional expressions of hope of learning something by exploring speculative ideas. It happens.

For the purpose of this book I name a third tradition for thinkers who, for many and various reasons, doubt what scientists, as pragmatists, have been saying about their theories. Let it be the skepticism camp, to include Schiller at the turn of the century, Kuhn in 1970, and Latour and Woolgar a decade after that. The Stanford

16. It is obvious, but must be said, that I am no philosopher. The traditions I mention are shaped around the practices of natural sciences in general, and in particular the sciences that are simple enough to allow displays of tight connections between theory and practice. I assign names that have some historical basis, but Menand (2001) and Misak (2013) can set you straight on the rich varieties of philosophies of science and society.

Encyclopedia of Philosophy entry on Skepticism (Comesaña and Klen 2019) describes a broad variety of thinking in the traditions of philosophical skepticism. I use the term, skepticism camp, simply to include all those who are not persuaded by the thesis I am presenting in this book, that the simpler physical sciences make a case for objective reality that is hard to ignore, persuasive though never a proof.

Peirce fits one end of the spectrum of thinking in the pragmatism camp, James and Dewey the other, or maybe one end of the spectrum of skepticism with relativism on the other. The Stanford Encyclopedia of Philosophy (Baghramian and Carter 2121) informs us that the philosophy of relativism,

> roughly put, is the view that truth and falsity, right and wrong, standards of reasoning, and procedures of justification are products of differing conventions and frameworks of assessment and that their authority is confined to the context giving rise to them.

This relativism is a reasonable description of the state of physical cosmology in general in 1960, and of thinking about the cosmological constant in 1990. But as the tests improved, the issues changed to those characteristic of the more mature theories with the predictive power treasured by those of us on Peirce's side of the spectrum of pragmatism.

Another useful classification places on one side the empirical constructionism practiced by Peirce and on the other the social constructionism advocated by Latour and Woolgar. The former certainly includes community thinking among natural scientists, along with the search for predictive power. Relativism is a limiting case of the social constructionism camp that has it that natural science is based on agreed-upon observations interpreted in terms of theories imposed by an elite class aided by a herd willingness to ignore inconvenient facts. Kuhn (1970) points out historical examples where things of this sort happened in science. I shall discuss in Sections 3.2 and 6.9 examples drawn from developments in physical science in the past century.

The arguments for social constructionism that flourished in some circles in the late twentieth century, with the challenge to physical scientists to consider this aspect of their research, did catch the attention of working scientists. They did not always grasp the lessons to be drawn from such thinking, and those in the social constructionism camp did not always demonstrate a firm grasp of what was happening in the empirical constructionism camp of the physical scientists. I aim to address the misunderstandings on both sides of this "science war" in Sections 2.2 and 7.1.

Another impression to be corrected is illustrated by Mach's complaint that "The real fundamental facts are replaced by an equally large number of hypotheses, which is certainly no gain." Physicists do feel free to adjust theories and their parameters in the search for theories that better fit the observations. One might ask, with Mach, whether they are making up just so stories, as in Kipling's (1902) *Just So Stories for Little Children*, on such things as how the leopard got his spots. The story helps us remember that leopards have spots, and so might be the case of a physical theory that is adjusted to fit the observations.

Along with the prospect of this circular construction of theories we must consider that, if we trust the evidence to guide us to the theory, we cannot be sure there is a single basin of attraction. If more than one theory continues to fit the evidence, what do we make of the claim to be succeeding in the pursuit of objective reality?

These complaints from schools in the skepticism or social constructionism camp could be read to mean that the physical sciences are on unsteady ground, but that is not so. Let us recall the situation in physical science a century ago.

---

The Case for Objective Reality a Century Ago

- Mach discussed how Newtonian physics accounts for phenomena ranging from the motions of the planets and the moons around them to the shape of the rotating earth, ocean tides and trade winds, the rotation of the plane of Foucault's pendulum, and the working of pendulum clocks.
- Peirce discussed how applications of Newtonian physics to astronomical observations determine our distance to the sun, and from that measurements of the speed of light, with results consistent with laboratory measurements and what is inferred from the study of electric and magnetic fields.

The physical science of 1900 included many facts on the ground, and many successful predictions of far broader ranges of phenomena than went into constructions of the standard theories of the time. Mach did not make much of this, and others have not been persuaded by the predictive power of physical theories, but the case is there and in my opinion difficult to ignore.

The philosopher Hilary Putnam (1982) later put the situation in a memorable way:

> The positive argument for realism is that it is the only philosophy that doesn't make the success of science a miracle.

I take the liberty of naming this argument Putnam's miracles. It is an accurate statement of the mindset of most people active in the physical science community, but I must note that Putnam and other philosophers are said to consider the statement a little simplistic. A nuanced review is presented in Section 2.1 in the Stanford Encyclopedia of Philosophy entry on Scientific Realism (Chakravartty 2017). In particular, we have no measure of the probability that there is some other theory that passes other tests that produce an equally satisfactory picture of the world around us. In natural science we simply rely on the intuition that this seems exceedingly unlikely. It is the best science can do.

The corruption of physical science by social pressure and over-enthusiastic promotion of theories is a real phenomenon; I will offer examples from the development of physical cosmology. But the commonsense conclusion from consideration of Putnam's miracles applied to the abundance of evidence that the predictive power of physical science has given us is that the corruption is marginal, and that a century ago there was a persuasive case that the physical theories then were useful approximations to objective reality. The case has since grown even more persuasive; it is summarized in Section 7.6.

We must remember, alongside this abundance of evidence to check ideas, two limitations to the scientific method in the empirical constructionism camp. First, our best physical theories are incomplete. Physicists at the turn to the twentieth century were uncomfortable with Thomson Baron Kelvin's (1901) two "clouds over the dynamical theory of heat and light." Though the predictive power of Newtonian mechanics was impressive, as it still is despite the even more impressive general theory of relativity, its application to statistical mechanics fails to account for measurements of heat capacities. This is Kelvin's Cloud No. II. He suspected that it is the result of the failure of materials to relax to true thermal (statistical) equilibrium. The remedy has been found instead to be quantum physics. Kelvin had good reason to be impressed by the theory of electromagnetism, which figures in the second entry in the box, because he owed his peerage and fortune to patents for transoceanic telegraph cables. But despite the predictive power of electromagnetism as Kelvin applied it, the theory as it was then understood fails to account for experimental investigations of the propagation of light, most famously Michelson's lack of detection of motion relative to the ether. Kelvin mentioned

> a brilliant suggestion made independently by FitzGerald and by Lorentz of Leyden, to the effect that the motion of ether through matter may slightly alter its linear dimensions, [but] I am afraid we must still regard Cloud No. I. as very dense.

The Lorentz-FitzGerald contraction Kelvin mentioned was a step from Cloud No. I to relativity physics. Though Kelvin's two clouds were resolved by the introductions of relativity and quantum physics, they were replaced by others. A particularly dense cloud now is the quantum vacuum energy density discussed in Section 6.3.

The claim of solid evidence in support of theories that are manifestly incomplete is not inconsistent, but it is awkward to explain. This was true a century ago and remains so. Maybe the perception of physical science by those outside the science community would have been better served by clearer advertisements of the clouds as well as the impressive ability to avoid having to appeal to ridiculous numbers of Putnam's miracles.

The second limitation to the use of physical science to support the argument for objective reality is that the celebrated successes are drawn from carefully controlled experiments or observational situations that can be interpreted in terms of reliable predictions by our fundamental theories. We are basing grand conclusions on special situations. We cannot reduce the results of research in molecular biology to the fundamentals of quantum physics, for example. The fact of reproducibility of experimental results and practical applications of molecular biology is encouraging, but the situation is too complicated for claims for or against the evidence of an objective reality that can be derived from our canonical physics. This discussion continues in Section 2.3.

Positivism took Mach a long way, but pragmatism led Peirce to ask the productive question, "how are we ever going to find out anything more" about hypothetical ideas? The pragmatic tradition practiced in physical science over the past century takes it that speculative ideas might produce testable predictions that either falsify the ideas or encourage further explorations and tests. It is difficult to find physical scientists who have spoken out for pragmatism; they are instead occupied in research in this tradition. But we have a good example in the chapter, "Against Philosophy", in Weinberg's (1992) book, *Dreams of a Final Theory*. The book title expresses Weinberg's vision, that objective reality is there to be discovered.

## 1.4   The Working Assumptions of Physics

We see during the past century traditions of doubt in the skepticism camp about whether the pragmatist physicists know what they are doing, and the pragmatist's tradition of ignoring those in the positivism and skepticism camps while producing advances in science and technology that are hard to dismiss as Putnam's miracles. I present here a statement of the central idea of the pragmatism philosophy that guides the practice of research in natural science, what I take to be the implicit working assumptions. It is motivated by two key observations.

First, centuries of experience have shown that experiments conducted in conditions that can be repeated yield repeatable results. This has been known as a practical matter at least since people started using spears and arrows, or maybe even earlier from the experience that piles of rocks stay put unless something observable disturbs them. Peirce and Mach were quite aware of this repeatability. I take it to be the empirical basis for the starting assumption for natural science, that objective reality operates in what we would term a rational way.

Second, centuries of experience have shown that the standard physical theories predict a far broader range of phenomena than went into their formulation. Peirce and Mach knew this perfectly well; examples of the evidence at their disposal are summarized in the box on page 39. Peirce recognized the implication: it is what would be expected of theories that are close enough to reality that they successfully approximate the way reality expresses itself in new situations. It is the empirical basis for the idea that objective reality operates by rules that we are discovering.

These two observations from experience motivate the form of pragmatism that is implicit in the practice of research in natural science. I make this practice explicit by the statements of four fundamental working assumptions.

---

Working Assumptions of Research in the Natural Sciences

1. Objective reality, or Nature, operates by our rules of logic and mathematics expressed in the approximations of physical theories discovered by the interaction of theory and observation.
2. Theories worth close attention are good enough approximations to reality that their applications yield successful predictions of situations beyond the evidence from which the theories were constructed.
3. Falsified predictions force adjustments of theories, or the creation of new ones, some of which will be found to be improvements by the test of predictions.
4. The iterative improvements of physical theories driven by empirical evidence will converge to a unique end, the nature of reality.

---

Bear in mind that these statements are assumptions. They need not be justified by anything other than the fact that they are found to be useful, but I have tried to illustrate the history of thinking behind them. Recall Peirce's (1877) proposal quoted on page 9, that there "are real things, whose characters are entirely independent of our opinions about them."

Peirce's assumption, and the assumptions in the box, require commentaries.

1. Some question the meaning of an objective mind-independent physical reality. Others take reality to be as

obvious as the chair I am sitting in. Working scientists seldom if ever pause to think about it, but they operate as if their research is guiding them in the direction of a reality to be discovered. I refer to the Stanford Encyclopedia of Philosophy entry in Scientific Realism (Chakravartty 2017) for the philosophical literature on how to specify reality. For our purpose it is best left as the assumption that research in the natural sciences is discovering approximations to objective reality.

2. It may seem obvious that reality, assuming it exists, respects our ideas of logic and mathematics. But since we have not been issued a guarantee, this also is an assumption, a hope.

3. The second and third points in the box assume empirical tests can convincingly falsify some predictions and demonstrate that others are consistent with the tests: measurements or observations. A test that is found to be close to but apparently significantly different from the prediction drives searches for systematic errors in the test and for room for adjustment of the theory. Accepting the latter calls for another round of tests of predictions. This approach has worked well so far in the simpler applications of quantum and relativity physics, such as cosmology, and the great progress in studying the physics of complex living matter suggests it is not a problem there either.

4. The fourth point in the box is an expression of hope in two directions. The first is that it will not be established that no theory fits all the empirical evidence, or that the evidence offers basins of attraction to more than one empirically satisfactory theory. The second is motivated by the experience of setbacks in science. Examples of an apparent success that proves to be a failure, and an apparent falsification that is found to be wrong, are considered in Chapter 4, on Einstein's argument that a logically satisfactory universe is close to the same everywhere. It illustrates the need to state the assumption that better theories will be found that build on earlier ones in an approach to reality, whatever that is.

5. Peirce wrote of "truth and reality" that would be "ultimately agreed to by all who investigate," but given his perception that nothing is precisely true, as we see in the quotation on page 13, he might have agreed that "truth and reality" might be asymptotically approached but never reached. The physical science community remains divided about the prospects of discovering ultimate reality, a theory of everything.

6. The standard interpretation of quantum physics requires a profound adjustment of our worldview: repeatability is to be understood in the sense of well-defined predictions of statistical patterns obtained by repeated measurements. But this is an adjustment of our understanding of what research is telling us about the nature of reality, not an adjustment of the working assumptions.

7. Our sense of elegance and beauty has an influential place in assessments of theories and designs of observations. This social side of natural science is real but cannot be entered as a starting assumption, because tastes for elegance are negotiable, adaptable to what works.

8. Another social force, intuition, can lead us to aspects of reality. An example is Einstein's argument from pure thought for a homogeneous universe. The thought passes abundant tests. Intuition does not belong in the working assumptions, however, because it can mislead. Consider Einstein's rejection of the cosmological constant that he introduced, and came to regret, but the tests of predictions of physical cosmology now require.

9. The empirical construction of reality will stall at what the world economy can afford to test, or maybe before that by the complexity of testing the most promising theories we can come up with. Physicists will venture further, and may be expected to arrive at what the community will agree surely is the final theory of everything. It will be a social construction.

10. Since human minds are constructing our physical theories, why speak of objective rather than subjective reality? Recall the repeatability of experiments and the predictive power of theories. I have mentioned examples from a

century ago, more to be discussed in this book are drawn from the relativistic theory of physical cosmology, and a still broader variety is to be found in other branches of physical science. They are the basis for the argument that our well-tested theories are useful approximations to the operation of a lawfully operating objective reality. We cannot do better; there cannot be a proof of objective reality from natural science.

11. We cannot predict the worldview of an organization of some sort on another planet, maybe in another planetary system, that takes an interest in such things. There is the exceedingly remote but perhaps real possibility someday to check whether such an organization arrived at something to compare to our working assumptions.

Two academic comments are to be added for completeness.

1. Repeatability is not to be confused with determinism, which already is falsified in classical physics by exponential insensitivity to initial conditions. Maybe this has something to do with free will, but we are concerned with far simpler issues.

2. Our fundamental physical theories are formulated in the compact forms of actions for the variational principle. An example is presented in footnote 2 on page 63. Other applications range from general relativity to the standard model for particle physics, and on to the search for a still better fundamental theory. But the working assumptions are best confined to the general use of mathematics, even though variational principles have been remarkably productive so far.

# CHAPTER TWO

# The Social Nature
# of Physics

Since society creates science, the progress of natural science is influenced by the broader culture of society. The point is more clearly recognized by sociologists than scientists. Sociologists form their impressions of natural science from what they observe scientists doing and what they are saying they are doing, and sociologists can see things that are not so obvious to those heavily involved in the science. I consider here two particularly relevant examples. One is the common occurrence of discoveries introduced independently by two or more people or groups. The sociologist Robert King Merton used the term multiples for this phenomenon, along with the singletons that have an arguably unique provenance. The second example from sociology is the theories that are accepted in the canons of physics but are only loosely tied down by empirical evidence. Sociologists have a term for these theories: socially established. Sociologists also may arrive at conclusions that physicists consider are not quite right, of course. It is the nature of academia, and examples will be considered.

## 2.1 Multiples

I expect anyone who has spent much time in research in physical science has noticed that, when an attractive idea appears, there is a reasonable chance that the idea had already been proposed, independently, or else will be independently proposed if news does not travel fast enough. I suppose the same situation is true in other branches of natural science, and for similar reasons. The trigger might be an advance in technology, or it might trace back to the way people communicate, not always directly or even verbally. I have never encountered a physicist who remarked that this situation surely is more interesting than simply an operating condition for our subject. Sociologists recognize it as a phenomenon that might teach us something. Following the sociologist Robert Merton's (1961) article, "Singletons and Multiples in Scientific Discovery: A Chapter in the Sociology of Science," I term this phenomenon in science and society Merton multiples.

The phenomenon is common. The sociologists William F. Ogburn and Dorothy S. Thomas (1922) published a list of 148 examples "collected from histories of astronomy, mathematics, chemistry, physics, electricity, physiology, biology, psychology and practical mechanical invention." They mentioned contributions to the special theory of relativity, but not to the general theory.

Ogburn and Thomas invited us to consider,

> Are inventions inevitable? If the various inventors had died in infancy, would not the inventions have been made and would not cultural progress have gone on without much delay?

In Merton's (1961) words, Ogburn and Thomas

> concluded that the innovations became virtually inevitable as certain kinds of knowledge accumulated in the cultural heritage and as social developments directed the attention of investigators to particular problems. . . . Appropriately enough, this is an hypothesis confirmed by its own history. (Almost, as we shall see, it is a Shakespearian play within a play.) For this idea of the sociological significance of multiple independent discoveries and inventions has been periodically rediscovered over a span of centuries.

Ogburn and Thomas put it that

> the elements of the material culture at any one time have a good deal to do with determining the nature of the particular inventions that are made.

Once the phenomenon of multiples has been pointed out we readily understand that, when the material culture allows a discovery, more than one group can take advantage of the opportunity and produce a multiple. An example to be discussed in Chapter 5 is the post-World War II contributions to cosmology by four influential physical scientists, each of whom independently decided to direct their research to aspects of gravity physics and cosmology. That is to say, their decisions were independent apart from the great release of energy for peacetime research in science and technology at the end of the war, and the inviting opportunity to add physical considerations to the bloodless prewar thinking about gravity physics and the large-scale nature of the universe.

The influence of ideas is familiar in everyday life, not surprising to see in the practice of physical science, and surely a contributing factor to the multiples phenomenon. An example to be discussed in Section 6.6 is the idea that one or more of the neutrinos of elementary particle physics may have a nonzero rest mass. That means the mass density in these neutrinos might be interesting for cosmology. The idea evolved through two sets of Merton multiples in the 1970s. I imagine these multiples were not purely accidental, but rather ideas that had been floating through the particle physics community. They were important, the starting point for the dark matter of modern physical cosmology.

Ogburn and Thomas (1922) pointed out that the introduction of wireless telegraphy may be expected to have "cut short the researches of other inventors along similar lines," thus reducing incidences of multiples. But the even more rapid means of communication in use since then have allowed development of a considerable variety of multiples in the constructions of physical theories. One might even imagine that the ease of broadcasting thoughts through the internet as they occur would encourage the spread of vaguely formed ideas that eventually reach people prepared to develop whatever aspects of the thoughts came through. The examples from physical cosmology discussed in Section 7.3 make the phenomenon of multiples clear and, I will argue, help us understand the interactions of science and society.

## *2.2  Constructions*

Following the practice begun on page 38, I use the label, "empirical constructions," for empirically tested concepts of objects in physical science and the ideas or theories by which scientists expect them to behave. Empirical constructions are expected to be likely to be accurate (though some say that if you're not occasionally wrong you're not trying hard enough). Sociologists, historians, and philosophers form social constructions of developments in physical science from what they observe scientists doing. Scientists form their own social construction when a theory looks too good to be wrong and enthusiasm runs ahead of empirical evidence. Social constructions may be found to be accurate, though not as often as empirical ones. Then there are circular constructions, theories that fit given evidence because they were designed to fit the evidence. They also are known as just so stories, of the kind discussed on page 39. Circular constructions are not unusual in the natural sciences; they are part of the process of search, invention, test, and then maybe move on.[1]

I take as a prototype for a moderately contentious social construction of physical science the report in the book *Laboratory Life*, by Bruno Latour and Steve Woolgar (in two editions, 1979 and 1986). It tells the interesting story of what the philosopher Latour made of his embedding for two years in the Salk Institute for Biological Studies. Latour is not a scientist, he had no prior instruction in the practice of this art, and he did not seek direct instruction while at the Salk Institute. In the introduction to *Laboratory Life*, Jonas Salk wrote that Latour acted as

> a kind of anthropological probe to study a scientific "culture"—to follow in every detail what the scientists do and how and what they think. He has cast what he observed into his own concepts and terms, which are essentially foreign to scientists. He has translated the bits of information into his own program and into the code of this profession. He has tried to observe scientists with the same cold and unblinking eye

---

1. I find no difficulty in distinguishing the three kinds of constructions in the topics in physical science discussed in this book. The physical science of living matter is more complicated.

with which cells, or hormones, or chemical reactions are studied—a process which may evoke an uneasy feeling on the part of scientists who are unaccustomed to having themselves analyzed from such a vantage point.

Salk offered no objection to Latour's observations, and I learned things from the Latour and Woolgar book. But Latour's observation of the absence of independent verification of results from the molecular biology laboratory (which is mentioned on page 33) misses the point Peirce made so well in his comments (discussed on page 6) about the consistency of independent ways to measure the speed of light. This is not surprising. Surely an anthropologist embedded in a culture that had managed to stay well isolated would arrive at an accurate grasp of some aspects of the culture while quite misreading others.

In the preface to the second edition of their book, Latour and Woolgar pointed out that the phrase, social construction, which they used to describe what Latour observed, is redundant because members of society are responsible for all constructions in natural science. The first edition of their book has the subtitle, *The Social Construction of Scientific Facts.* Along with making this point in the second edition, Latour and Woolgar changed the subtitle to *The Construction of Scientific Facts.* But for our purpose both terms, social and empirical constructions, are useful.

An example along similar lines is Karin Knorr-Cetina's (1981) report, in the book *The Manufacture of Knowledge: An Essay on the Constructivist and Contextual Nature of Science,* on her observations "from October 1976 through October 1977 at a government-financed research centre in Berkeley, California." She states that her

thesis under consideration is that the products of science are contextually specific constructions which bear the mark of the situational contingency and interest structure of the process by which they are generated, and which cannot be adequately understood without an analysis of their construction.... My observations focused on plant protein research, an area which turned out to include aspects of protein generation and recovery, purification, particle structure, texture, assessment of biological value, and applications in the area of human nutrition.

This combines observations of the pure curiosity-driven study of complex systems with considerations of the social consequences of the research, from biophysics to nutrition. It looks like a rich subject for study by a sociologist. But again, the complexity prevents discovery of evidence for or against the case that this science rests on an objective reality that is usefully approximated by the predictions of our accepted theoretical physics.

Other sociologists, including Andrew Pickering in the book, *Constructing Quarks* (1984), have reported observations of research in elementary particle physics at particle accelerator laboratories. On the face of it this too is complicated, requiring the organization of large groups of scientists and engineers to design, construct, and use particle accelerators and detectors in pursuit of the standard model of particle physics. The physical theory that grew out of this research is elegant, in the sense Frank Wilczek (2015) describes in *A Beautiful Question: Finding Nature's Deep Design*, though much more complicated than Newtonian physics. The theory has 26 parameters, depending how you count, which is a lot. The tests require the guidance of the theory to sort through the spray of particles produced in an energetic particle collision, which is circular. The theory is incomplete, with no place for dark matter and the cosmological constant, for example. But let us not be distracted from the important point. Elementary particle theory now is well specified and many of its predictions are definite and reliably computed. The predictions fit a wealth of well-checked reproducible tests. Could a different theory of the structure of matter fit a comparable range of empirical evidence, maybe obtained by tools other than all those that have been used to probe matter and arrive at the standard theory? The same question applies to all our natural sciences, of course. A falsification of the idea is not possible, but we can observe that if two different theories of matter passed all the tests from laboratory and particle accelerator experiments, the interpretation would require an array of coincidences that seems absurd, a variety of Putnam's miracles argument.

I take as prototype for a seriously contentious argument for social constructions David Bloor's (1976) assessment of science from the standpoint of the "strong programme of sociology," expressed in the book *Knowledge and Social Imagery*:

knowledge for the sociologist is whatever men take to be knowledge. It consists of those beliefs which men confidently hold to and live by. In particular the sociologist will be concerned with beliefs which are taken for granted or institutionalised, or invested with authority by groups of men.

This is a durable way of thinking; Schiller was expressing similar thoughts (quoted on page 13) a century ago. It is an accurate description of some aspects of natural science. For example, in 1960 Einstein's general theory of relativity was "invested with authority" by accomplished and respected physicists who did "confidently hold to" general relativity for reasons that they considered persuasive, but were almost entirely social. The situation is discussed in Section 3.2. I am not complaining about the acceptance of general relativity in 1960; I am offering an accurate statement of the situation.

In the foreword to the second edition of his book, Bloor (1991) added this comment:

> But doesn't the strong programme say that knowledge is purely social? Isn't that what the epithet 'strong' means ? No. The strong programme says that the social component always is present and always constitutive of knowledge. It does not say that it is the *only* component, or that it is the component that must necessarily be located as the trigger of any and every change: it can be a background condition.

Indeed, scientific research has a social component even though scientists seldom bother to think about it. But we must consider both sides of research in natural science: the social influences and the pursuit of empirical tests of predictions.

In the year 1900 physical scientists confidently held to electromagnetism and Newtonian physics for a good reason: the broad predictive power that is reviewed in Chapter 1. We have seen (on page 9) Peirce's discussion of the consistent results of four quite different ways to measure the speed of light. They required different instrumentation, and they depended on auxiliary assumptions: the theory of motion of the planets for some, experiments with electricity and magnetism for another. The consistency remains an impressive example of the successful confrontation of theory and observation. There were other examples a century ago; consider the electrification of communities. If standard physics had failed to account for the generation, transmission, and consumption of electromagnetic energy we would have heard about it. We have many more examples now: we confidently hold to relativity and quantum physics because of the even greater breadth of their well-tested predictive power. All this is serious empirical evidence, by the criterion of Putnam's miracles, that we are approaching objective reality.

For Bloor, the scientists' case that this is so would imply that

> The sociology of knowledge is thus itself unworthy of belief or it must make exceptions for scientific or objective investigations and hence confine itself to the sociology of error.

There is ample room for sociologists' analyses of what physicists do, particularly their mistakes. The many results of "scientific or objective investigations," the tests of predictions and the products of this research that have contributed so much to society, for better or worse, must be considered too. How could we ignore this broad array of evidence?

Thinking in another direction is presented by Richard Dawid (2013) in the book, *String Theory and the Scientific Method*. Dawid applies the term, "non-empirical confirmation," to the acceptance by the physics community of social constructions that are inspired by generalizations from well-founded empirical constructions. Dawid's prime example is the community fascination with the compelling elegance of superstring theory. The concept follows in a natural way from the quantum field theory of elementary particles, a theory that passes multiple demanding tests. The new ideas of variants of superstring theory offer the promise of empirical contact, but that awaits a theory definite enough for testable predictions. For our purpose another, earlier, example of non-empirical confirmation is more directly relevant.

The theory of the electromagnetic field was well tested and widely applied, and already relativistic, before Einstein showed us why the special theory of relativity ought to apply to the rest of physics. Einstein's general relativity, another field theory, follows in an elegant way by precepts drawn from the field theory of electromagnetism. This is easier to say now, of course, but the point is that the general theory of relativity began as an accepted social construction, non-empirically confirmed by its admirable pedigree. The theory only later graduated to a well-tested empirical construction. (I emphasize the warning label: this successful transition does not imply a guarantee of empirical confirmations of other non-empirically confirmed theories.)

The discussions of constructions, or paradigm additions, to be considered in the following chapters are meant to illustrate four points. First, constructions may become canons of physics because they have passed searching empirical tests, or on occasion because their relations to other standard and accepted physics that has passed demanding tests are too elegant to resist. The latter are social constructions, but with the important qualifier that they are motivated by examples of productive empirical constructions. They are non-empirically confirmed; the cases are not persuasively established; but they appear to be worthy of close attention.

The second point is that at least two kinds of Merton multiples produce constructions. Some are the clear results of social pressures and the opportunities afforded by developments of technology. Others bring to mind more subtle, perhaps nonverbal, communications, as thoughts that are somehow "in the air," or these days floating in the internet cloud.

Third, a social construction may be found to offer predictions that pass empirical tests. As mentioned, this happened to general relativity. When it happens the construction is promoted to an empirical construction because empirical evidence has reenforced the case that the construction is a useful approximation to what we are conjecturing is objective reality.

Fourth, though I do not mean to underestimate the role of society in determining the way we behave in science, I do not want to obscure the role of observational and experimental evidence in motivating constructions and determining which are more or less convincingly established approximations to reality.

The paradigm additions in the building of physical cosmology, the example to be examined in Chapter 6, were products of the usual mix of assessments of evidence influenced by ideas. The distinguished astrophysicist Geoffrey Burbidge liked to term the very real influence of the natural science community acceptance of ideas that look interesting the "bandwagon effect." Burbidge emphasized that the effect can be problematic, encouraging reluctance to reconsider ideas that a broader assessment would reveal do not look promising. But the bandwagon effect also can be useful, concentrating attention to specific ideas until examination of the empirical evidence compels acceptance or adjustment. It reduces the chaos of many ideas on the table. Either way, bandwagons have a real effect on science.

For the purpose of exploring the nature of natural science, the advantage of considering the history of paradigm building in the relativistic theory of physical cosmology is the simplicity of the phenomena and their interpretation within well-defined and relatively simple laws of physics. This allows a clear view of the working assumptions of natural science. My assessment is listed on page 43. It biases the conclusions because theory and practice are more complicated in most of physical science. Perhaps phenomena ranging from chemistry and on up in degrees of complexity to living matter, when better understood, will lead us to more interesting theories, or better working assumptions, or maybe to confusion. Let us consider some aspects of this last thought.

## 2.3 The Sociology of Scientific Knowledge

Since research in physical science is conducted by members of society it is natural, once physicists face up to it, for them to agree that scientific conclusions are conditioned by society. Perhaps it is natural to follow that up by asking whether results from research in physical science are in some significant measure social constructions rather than probes of reality. In Harriet Zuckerman's (2018) words, in the study of the

> Sociology of Scientific Knowledge [known as SSK] ... the real challenge these publications declared was to demonstrate how society and culture determined the substance of knowledge claims scientists made. This new agenda was not to supplement the institutional approach [associated with Merton and Zuckerman], it was to be seen as the only way to understand how science worked. The criticisms to which constructionists gave the strongest weight concerned "deep flaws" in the

institutional approach owing to its unexamined positivist assumptions and its reliance on the role of social norms played in regulating the behavior of scientists.

Zuckerman is closely associated with Merton. She recalls that Merton

> was known to say that he would prefer to fly on an airplane designed in accord with scientific definitions of aerodynamics than on a plane built as a social construction.

Here is how a seasoned physical scientist, Peter Saulson (in Labinger and Collins 2001, page 287), reacts to the sociology of scientific knowledge, SSK.

> I propose the following thought experiment: ask anyone with a position of responsibility in our industrial economy whether she would care to do her job using only technology uninfluenced by scientific discoveries of the last seventy-five years. My prediction is that you will get few takers. What does this prove? ... A theory that is more powerful (in the sense of describing the world better) will be preferred by those with absolutely no cultural commitment to the content, expression, or metaphysical baggage of the theory. ... The global positioning system (GPS) has become ubiquitous in the military world for which it was designed as well as in such benign activities as hiking in wilderness areas. The precise time keeping on which GPS depends is sensitive to a tiny [general] relativistic effect, that the rate at which a clock registers the passage of time depends upon the gravitational field in which the clock sits. The orbiting clocks on the GPS satellites wouldn't give the answers on which the system depends unless their special gravitational environment were taken into account.

Debates over SSK, and the issue of social versus empirical constructions, were heated enough to be termed the science wars. Arguments that physical science is probing the inherent properties of reality, which I am presenting in this book, focus on controlled experiments with reproducible results that can be interpreted within the predictions of well-tested theoretical physics. The results encourage what sometimes may be overconfident pronouncements of the successes of science. Thinking on the other side, social constructionism, may have been encouraged by that overconfidence, and by scientists' tendency to make up fables or, more charitably put, to argue for positions that others can see are seriously problematic. Collins and Pinch (1993) present examples in their book, *The Golems: What Everyone Should Know about Science*. Among their examples are serious aberrations that some scientists took more seriously than they should. This sort of thing gets corrected, but meanwhile can sow confusion.

We have seen in Section 2.2 another factor. The reports by sociologists of their observations of studies of the physics of living matter have been influential on the social constructionist side. This research calls for a fascinating variety of technologies that probe deeply complex situations, and it requires speculative ideas on how to interpret the results. I admire this research but can imagine that it might not inspire the confidence of an onlooker who does not have practical experience in the ways of science. In this line of research it is not possible to make the neat connections of theory and observation that allow clear tests of predictions of relativity and quantum physics. It is important that controlled experiments on these complex systems are reproducible, and that the research has led to practical and effective applications, which also indicates reproducibility. These are facts on the ground, an important signature of the concept of a rationally operating reality. But they make a less than persuasive case that we have the evidence that our starting assumptions (listed on page 43) would lead us to expect of a useful approximation to objective reality.

The difference between the simplicity of gravity physics and the complexity of biophysical molecules is large but the divide is not sharp; natural science deals with all varieties and degrees of complications. But let us turn to an assessment of the case for reality on the simple side, gravity physics.

# CHAPTER THREE

# The General Theory
# of Relativity

Einstein arrived at his general theory of relativity by carefully thought through considerations that owed as much to his intuition as to empirical evidence. By 1950 the elegance of Einstein's theory had earned it a place in the canons of physics, but the empirical basis was not much better than when Einstein discovered his theory in 1915. The theory was a social construction. The precision tests that began in 1960 gave us a compelling case that the theory is an excellent approximation to the behavior of gravity on scales ranging from the laboratory to the solar system, and on to the well-tested theory of the large-scale nature of the universe. This chapter deals with the progress from discovery to precision tests. The role of general relativity in establishing that our universe is evolving from a dense hot earlier state is the subject of the rest of this book.

## 3.1   The Discovery

An authoritative source for the story of how Einstein arrived at his general theory of relativity is *The Genesis of General Relativity*, in four volumes, edited by Jürgen Renn (2007). Briefer accounts are in Janssen and Renn (2015) and Gutfreund and Renn (2017). We may draw from these sources the following eight considerations that guided Einstein's search for a relativistic theory of gravity, and led him to be satisfied with his result in 1915.

1. The general theory of relativity accounts for the insensitivity of gravitational acceleration to the natures of the test particles.
2. The theory reduces to Newtonian gravity in the limit of small speeds, where we know Newton's theory is a good approximation.
3. It accounts for the discrepancy in the Newtonian theory of the motion of the planet Mercury.
4. It is generally covariant: labels of positions in spacetime are only computational tools.
5. It replaces Newton's concept of an absolute spacetime with Mach's conditioning of spacetime by mass.
6. It satisfies local conservation of energy and momentum, as do all the other established physical theories.
7. It is a field theory constructed along the lines of electromagnetism.
8. It is elegant, a condition we cannot define but feel we know when we see it.

The first of these points refers to Einstein's great thought that the effects of gravity we experience are gravity-free behavior in curved spacetime. It would explain why the free fall acceleration of gravity is observed to be quite insensitive to the natures of test particles; they are just moving freely. This is known as the equivalence principle: what I experience as gravity pulling me down in my chair is equivalent to my acceleration in spacetime by the force exerted on me by my chair.

It is of course essential that a credible new theory of gravitation reduces to the Newtonian theory in the limit where it is known to work well, on Earth and in the solar system, apart from small relativistic corrections. In Einstein's general theory of relativity, Newton's theory is a good approximation when motions are much slower than the speed of light and spacetime curvature fluctuations are small. In the language of Newtonian gravity, the second of these conditions is that particles moving at nonrelativistic speeds through gravitational potentials acquire changes of speed that are much smaller than that of light. These are the conditions under which the nonrelativistic Newtonian theory is broadly and successfully applied.

During the decade that Einstein had been seeking a relativistic theory of gravity he knew the evidence that the motion of the planet Mercury departs from the Newtonian theory, and he felt that the better relativistic gravity theory he was seeking ought to account for the anomaly. Correspondence reveals his pleasure, we might say his relief, on finding that the 1915 theory properly accounts for the anomaly. This story continues in Section 3.3.1.

General invariance, or covariance, is the condition that the physical predictions of the field theory do not depend on how coordinates are assigned to positions in spacetime. In our present way of thinking this makes perfect sense: coordinate labels are only arbitrary assignments for the convenience of computation.

Einstein had expected that, in a logically complete theory, acceleration would have meaning only relative to the rest of matter, as Mach had argued and Einstein named Mach's principle. There is something to this in general relativity, because the theory predicts that inertial motion is determined in part by the motion of nearby matter. But, as Einstein came to appreciate the role of boundary conditions in determining inertial motion, he fell back on the postulate that a logically arranged universe would be the same everywhere, apart from local fluctuations. It would give inertia a universal nature consistent with the universal distribution of matter, arguably consistent with Mach's principle. The history of this postulate, which became known as Einstein's cosmological principle, is considered in Chapter 4. Einstein proves to have been right about the large-scale nature of the universe, but by the time we had interesting evidence of this he had lost interest in Mach's principle. We are left to speculate whether Einstein in 1917 was right about the universe for the right reason.

Energy and momentum are conserved in classical mechanics and electromagnetism. General relativity replaces this with local conservation laws that fit what is observed and remain part of standard physics. But general relativity does not offer a universal analog of the global conservation of energy and momentum that we have in classical physics. This interesting paradigm adjustment has passed the tests so far.

As noted in the first point, in general relativity our perception of the gravitational acceleration of a freely falling particle is equivalent to the motion of the particle without acceleration in curved spacetime. A measure of this curvature is derived from the metric tensor $g_{\mu\nu}$. This is a field, that is, a function of position in spacetime.[1] Einstein's field theory describes the behavior of the field $g_{\mu\nu}$ in response to the material contents of spacetime. In the same way, Maxwell's field theory describes the behavior of the electromagnetic field vector $A^{\mu}$ in response to the distribution and motion of electric charge. The indices $\mu$ and $\nu$ on the field $g_{\mu\nu}$ have the values 0, 1, 2, and 3, one component in the timelike direction, three in spacelike directions. The one index, $\mu$, on the electromagnetic field $A^{\mu}$ ranges over these four values. Classical electromagnetism is the simplest theory of conserved charge, if you want it to be covariant in the special or general theories. General relativity is the simplest relativistic covariant theory of gravity as curved spacetime with local conservation of energy and momentum. When Einstein introduced general relativity, and $g_{\mu\nu}$, electromagnetism with the field $A^{\mu}$ had already passed the test of experiments and a great many practical applications. In retrospect, at least, it made good sense to try applying the precepts of electromagnetism to the construction of a theory of gravitation. This is point seven. But we might pause to consider that the simplest of all possible relativistic theories of a homogeneous universe that is an acceptable home for us has no space curvature and no cosmological constant. Our instinct for simplicity has been confounded by the discovery that physical cosmology requires the presence of Einstein's cosmological constant, with a numerical value we do not understand. Simplicity can be a helpful guide, but we can't rely on it. Empirical tests can inspire confidence.

Point seven also includes David Hilbert's demonstration that Einstein's field equation can be expressed as a variational principle

---

1. The physical, coordinate-independent, separation $ds$ between two close events in spacetime at coordinate separation $dx^{\mu}$ is given by $ds^2 = \Sigma_{\mu,\nu} g_{\mu\nu} dx^{\mu} dx^{\nu}$. The indices $\mu$ and $\nu$ range over 0, 1, 2, and 3, for the components of position in time and space. Depending on the sign of $ds^2$, the square root of its absolute value is the distance between the positions of the two events measured by an observer moving so the events happen at the same time, or, with the other sign of $ds^2$, it is the time elapsed between the two events measured by an observer moving so the events happen at the same position.

with an elegantly compact and manifestly covariant action. The same is true of electromagnetism. This principle in the form usually applied in fundamental physics requires that the action computed from a solution to the field equation is not changed by small, infinitesimal, changes from the solution.[2]

On the face of it, this focus on the variational principle seems curious. Mach (1883, 1902) began his discussion of variational principles in mechanics and optics with the remark that

> MAUPERTUIS enunciated, in 1747, a principle which he called "*le principe de la moindre quantité d'action*," the principle of *least action*. He declared this principle to be one which eminently accorded with the wisdom of the Creator.

Maupertuis may have considered the word "least" to be of some theological significance, but it is a slight misnomer for the form of the principle usually applied in fundamental physics, because a solution only need be at a stationary point of the action (a maximum, minimum, or saddle point). Mach (1902) presented many applications of varieties of this principle to optics and static and dynamical mechanics.

The applications of the action principle to electromagnetism and general relativity are of lasting value. The principle also figures

---

2. For an example in classical mechanics consider a particle of mass $m$ moving in one dimension with potential energy $V(x)$. In a trial solution the position as a function of time is $x(t)$, the velocity of the particle is $dx/dt$, the Lagrangian is $L = m(dx/dt)^2/2 - V(x)$, which is the kinetic energy minus the potential energy, and the action is $S = \int dt\, L(x, t)$. The variational principle is that a solution to the equation of motion is at a stationary point of the action $S$, meaning $S$ is unchanged by a slight, infinitesimal, variation of the path of the particle. So consider changing the path along which the particle moves from $x(t)$ to $x(t) + \delta x(t)$, where the infinitesimal change is $\delta x(t)$. We can write the change in the action as

$$S[x + \delta x] - S[x] = \int dt \left[ m \frac{dx}{dt} \frac{\delta dx}{dt} - \frac{dV}{dx} \delta x \right] = - \int dt \left[ m \frac{d^2 x}{dt^2} + \frac{dV}{dx} \right] \delta x.$$

The last expression follows by integration by parts, where $\delta dx/dt = d\delta x/dt$, because partial derivatives can be taken in either order. Since the change in the action is required to vanish for any infinitesimal change in the path, the expression in brackets in the last term must vanish. The equation of motion thus is, as usual,

$$m \frac{d^2 x}{dt^2} = -\frac{dV}{dx}.$$

in the standard model for particle physics, Feynman's sum over histories formulation of quantum physics, and superstring models for a possible next big advance in fundamental physics. The role of this methodology in physics is important enough to be noted in commentary (2) on the working assumptions of physics listed on page 46 in Chapter 1.

Hilbert was prepared for his application of the action principle to general relativity. The last of Hilbert's (1900) storied 23 mathematical problems is "Further development of the methods of the calculus of variations."

One still encounters the statement that Hilbert independently discovered Einstein's field theory, maybe even before Einstein. But Einstein conceived the idea and the problem: find the field equation for curved spacetime with matter and the electromagnetic field. Hilbert's paper on the derivation of the field equation from the variational principle records the date of submission to the journal as five days before reception of Einstein's formulation of the equation. But Corry, Renn, and Stachel (1997) present a clear case, based on two different sets of proofs of Hilbert's paper, that the published version that presents the right field equation differs from the one submitted before Einstein's paper. Corry (1999) discusses Hilbert's thinking in more detail. It is a small matter, but Einstein got there first.

Einstein found the field equation for general relativity, and Hilbert the action for the variational principle, by brilliant strokes of intuition in physics and mathematics. Richard Feynman argued (in Feynman, Morinigo, and Wagner 1995) that Einstein's field equation follows in a less intuitively demanding way by the example of classical field theory, while demanding local conservation of energy and momentum and the proper Newtonian limiting case, and keeping the theory as simple as possible. This also is a demanding exercise.

I put considerations of elegance last on the list, in point eight. The concept is widely advertised, influential, but impossible to define because elegance is negotiable, adaptable. The elegance of general relativity might be said to be that the pieces of the theory fit together in a logical and simple way that parallels the successful theory of electromagnetism.

## 3.2   The Social Construction

In the 1950s Einstein's general theory of relativity was a standard and accepted part of theoretical physics. Problems on simple applications of the theory were part of the graduate general examinations I wrote in 1959 when I was a graduate student at Princeton University. The theory is presented in the book, *The Classical Theory of Fields*, which is part of the magnificent series of books on theoretical physics by Lev Landau and Evgeny Lifschitz. In my copy of the 1951 English translation of the 1948 Russian second edition of *The Classical Theory of Fields*, 202 pages are devoted to the classical theory of electromagnetism, the behavior of electric and magnetic fields in their interaction with the distribution and motion of electric charge. The last 107 pages present the general theory of relativity. Why were the two theories given roughly equal time?

Classical electromagnetism passes demanding tests, in the laboratory and in enormous varieties of practical applications, from the generation of electromagnetic energy in massive power plants to the energy transmission lines and down to the trickle of energy used to operate a cell phone. Classical electromagnetism is wrong in applications to small scales or large energies, meaning it is incomplete, but it belongs among the most thoroughly and persuasively established physical theories. The case for the general theory of relativity in the 1950s was far more modest, largely the list of considerations at the start of this chapter. This was persuasive to Einstein, and eventually to the physics community, to the point that general relativity had become part of standard theoretical physics by 1950. But the empirical case was weak, as will be discussed in Section 3.3.

In their series of books on theoretical physics Landau and Lifshitz offer scant mention of experimental evidence or explanations of why the elements of physical theory have been found to be what they say they are. Instead, for the most part the laws are simply stated in elegant compact ways and the many consequences systematically derived. The 1951 edition of *The Classical Theory of Fields* makes no mention of the great difference in the empirical support for the two classical field theories they present, electromagnetism and general relativity, and there is little mention of Einstein's physical considerations of how a relativistic theory of gravity is to be constructed.

Though the classical theory of electromagnetism was found before special relativity, Landau and Lifshitz start their exposition of electromagnetism from the special theory of relativity and the Lorentz covariant[3] action for the electromagnetic field. They approach the general theory of relativity by replacing the metric tensor that describes the flat spacetime of special relativity with a general function of position in spacetime, the field $g_{\mu\nu}(\vec{x}, t)$, and they derive the field equation for $g_{\mu\nu}(\vec{x}, t)$ from Hilbert's action. Their explanation for this choice of the action is minimal: it satisfies general covariance, and the field equation derived from the action contains the field and its first and second derivatives. There are precedents for this latter condition, from Newtonian gravity physics and electromagnetism, where the second derivatives give us the observed inverse square laws.

Earman and Glymour (1980a,b) document early confusion about the predictions of this new theory. The issues are subtle, despite the advertisement that this is the simplest of all possible relativistic field theories of gravity. But that was straightened out by the 1930s; the exposition is clear and standard in Richard Tolman's (1934) book, *Relativity, Thermodynamics, and Cosmology*. There were later additions to the analytic methods needed for the demonstrations of singularity theorems in general relativity, and the analytic and numerical methods needed for the computation of properties of relativistic stars and black holes, gravitational wave emission detected by pulsar timing, and detection of gravitational waves from merging black holes and compact stars. But the physical theory with the three classical tests is recognizable in Tolman (1934).

Let us consider now the state of these tests of general relativity in the post-World War II years, when Landau and Lifshitz had completed the second edition of their book on the classical theory of fields. These considerations are followed in Section 3.4 by

---

3. Recall that the physical predictions of a Lorentz covariant theory do not depend on how coordinate labels are applied to the flat spacetime of special relativity.

a discussion of the more demanding tests introduced beginning in 1960.

## 3.3 The Early Tests

Apart from the hint to a relativistic gravity theory from the examples of electromagnetism and special relativity, Einstein had just three reasonably reliable sets of observations that helped guide him to his general theory of relativity. First, he knew that the gravitational accelerations of small freely falling objects do not depend on the compositions of the objects, within the limits of precise measurements. This helped inspire his thought that what we see as gravitational acceleration is free motion without acceleration in curved spacetime. The idea is beautiful, but the experimental evidence of it cannot be counted as a check of a prediction because the evidence inspired the theory. (New, more precise tests of the universality of gravitational acceleration of test particles, and tests on scales much smaller and larger than could be reached a century ago, certainly count as checks of predictions.) Second, Newton's gravity theory was known to give a successful account of a broad range of astronomical observations and laboratory experiments. The choice of a candidate for a new theory of gravity that does at least this well was a condition, not a prediction. Third, Einstein knew the evidence that the orbit of the planet Mercury is significantly different from what is expected in Newton's gravity theory. Einstein's theory accounts for the difference. Prior to 1960 it was the one clear and unambiguously successful prediction of general relativity. This requires some discussion.

### 3.3.1 THE ORBIT OF MERCURY

In the nineteenth century, fitting the Newtonian theory to the observations of the motion of the planet Mercury required the postulate of unseen mass in the inner solar system, most conveniently placed in a hypothetical planet with mass comparable to that of Mercury, and in an orbit between Mercury and the sun. The new planet had a name, Vulcan, but increasingly sensitive searches failed to reveal it. In 1909 William Wallace Campbell, director of the Lick Observatory in California, reported that advances in astronomical photography had shown that an

> intramercurial planet could scarcely be larger than thirty miles in diameter and that roughly a million such bodies, of great density, would be required to supply the disturbing effect observed in Mercury's orbit.

Campbell (1909) mentioned the thought that the interplanetary dust that scatters sunlight to produce zodiacal light, the faint diffuse glow of light around the belt of the sky that includes the orbits of the planets, might be massive enough near the sun to have the wanted effect on Mercury's orbit. Willem de Sitter (1913) thought this could be possible, but Harold Jeffreys (1916, 1919) pointed out that it is exceedingly difficult to see how the wanted amount of mass could scatter the very little sunlight observed as zodiacal light. There also was talk of a possible departure from the exact inverse square law of Newtonian gravity, but this ad hoc adjustment was not considered a very interesting idea. In 1915 Einstein's general theory of relativity offered a resolution to a serious problem.

In 1907, while still working at the patent office in Bern, Einstein wrote to his friend Conrad Habicht[4] that

> At the moment I am working on a relativistic analysis of the law of gravitation by means of which I hope to explain the still unexplained secular changes in the perihelion of Mercury.... So far, however, it does not seem to be going anywhere.

An unpublished manuscript[5] in the handwriting of Einstein and his friend Michele Besso, on the analysis of the motion of the perihelion of Mercury, is dated largely to 1913. We see that Einstein continued to think about this anomaly. In a letter to Hendrik Lorentz, in January 1916, Einstein recalled that[6]

> Trying times awaited me last fall as the inaccuracy of the older gravitational field equations gradually dawned on me. I had already discovered earlier that Mercury"s perihelion motion had come out too small....

4. In the English translation in the Collected Papers of Albert Einstein, Vol. 5, Doc. 69

5. Discussed in the Collected Papers of Albert Einstein Vol. 4: The Swiss Years, pp. 344–359

6. Collected Papers of Albert Einstein Vol. 8, Doc. 177

> Now I am more pleased than ever about the arduously won lucidity and
> about the agreement with Mercury's perihelion motion.

Since Einstein was looking for a theory of gravity that would
account for the departure of the orbit of Mercury from Newtonian dynamics, his 1915 demonstration that his general theory of
relativity accounts for the anomaly was not a prediction, strictly
speaking. He was looking for a theory that passes this test, and
he stopped when he found one. But the seven other considerations
listed on page 59 offer a good case that, if Einstein had not known
about the Mercury anomaly, he would have been satisfied with his
1915 theory and stopped looking for something even better. That
is, it is reasonable to take it that the anomaly was a prediction
that passed observationally secure measurements. This is important, because until 1960 the other two classical tests of general
relativity, the gravitational redshift and the deflection of light by
mass, were far less secure.

### 3.3.2 GRAVITATIONAL REDSHIFT

General relativity predicts that the frequency of light received at
great distance from the surface of a massive body is shifted to the
red, to longer wavelength, by an amount determined by the mass
and radius of the body. (The fractional shift of the wavelength $\lambda$
of light leaving the surface of a spherical mass $m$ with radius $r$ is
$\delta\lambda/\lambda = Gm/rc^2$, to lowest order in the value of $Gm/rc^2$.) St. John
(1928) reported measurements of redshifts of absorption lines in
the spectrum of light from the sun. In a later review of the tests
of the special and general theories of relativity, St. John (1932)
concluded that

> The investigation confirms by its greater wealth of material and in
> greater detail ... that the causes of the differences at the center of the
> sun between solar and terrestrial wave-lengths are the slowing of the
> atomic clock in the sun according to Einstein's theory of general rela-
> tivity and conditions equivalent to radial velocities of moderate cosmic
> magnitude and in probable directions, whose effects vanish at the edge
> of the sun.

**Table 3.1.** Predicted and Measured Redshifts of Sirius B

|  | mass | radius | redshift predicted | measured |
|---|---|---|---|---|
| St. John 1932 | 0.85 | 0.028 | 19 | 19 |
| Greenstein et al. 1971 | 1.02 | 0.0078 | $83 \pm 3$ | $89 \pm 16$ |

units: solar mass, solar radius, km s$^{-1}$

Indeed, the majority of the lines are displaced to the red, many by about the predicted amount. But measurements of other line displacements differ for different strengths of lines of a given chemical element, for different elements, and at different places across the face of the sun. As St. John mentioned in the 1928 paper, the spectral line shifts are affected by Doppler shifts from the flow of plasma near the solar photosphere. There was no theory of these currents, no plausible way to correct for their effect. Added to this was the confusion from overlapping absorption lines. It was not a very reassuring test of gravity theory.

St. John (1932) also reported the application of the redshift test to the white dwarf star Sirius B. Its mass is comparable to that of the sun, and its much smaller radius makes the predicted gravitational redshift from the surface much larger. This white dwarf is gravitationally bound to the main sequence star Sirius A. The much more modest density of this star means its predicted gravitational redshift is much smaller than for Sirius B. This allows separation of the gravitational redshift of Sirius B from the effect of the Doppler shift due to its motion relative to us, because that motion is shared with Sirius A, whose Doppler shift can be measured. The first line in Table 3.1 shows St. John's report of data for Sirius B. The mass is based on a fairly reliable model for the mass of its companion, Sirius A, from its luminosity. Given that mass, the observed motions of the two stars around each other give the mass of Sirius B. The radius of Sirius B is based on its measured surface temperature and luminosity. (Given the surface temperature, and assuming thermodynamic equilibrium at the photosphere where the light becomes free of scattering and leaves the star, the temperature gives the rate of radiation of energy per unit surface area.

The product with the surface area, $4\pi r^2$, is the luminosity. Since the luminosity was measured, one could solve for the radius $r$.) The mass and radius give the predicted gravitational redshift entered in the first line of the table. The redshift is stated as the equivalent of the Doppler shift of motion away from an observer. The measured redshifts, from two different telescopes, gave consistent results, which happened to be the same as the prediction to two digits. St. John estimated the possible uncertainties might be 3 to 5 km s$^{-1}$.

The test looked good. But Gerard Kuiper (1941) reported an analysis that yielded a predicted gravitational redshift of about 30 km s$^{-1}$, well above the reported measurements. Kuiper cautioned that, with his estimate of the mass and radius, Chandrasekhar's (1935) model for the structure of a white dwarf star indicated that the hydrogen abundance by mass fraction in this white dwarf is $X \sim 0.5$. (The abundance of hydrogen matters because it affects the mean molecular weight.) Kuiper pointed out that this large value of $X$

> raises the serious difficulty of explaining the low *energy generation* in this White Dwarf having such a large concentration of protons packed to densities of $10^6$ and higher.

Kuiper did not explain, but he may have had in mind Wildhack's (1940) demonstration that, if there were a significant amount of hydrogen in the dense central regions of a white dwarf star, the energy released by the fusion of protons to make deuterons would far exceed the luminosities of white dwarf stars. Kuiper found that, if it were assumed that there is no hydrogen in the interior of Sirius B, then the best estimate of the predicted redshift would be 79.6 km s$^{-1}$, even larger than the reported measurements. Kuiper doubted this result, because it seemed to require an unacceptably large surface temperature, but it does agree with the models and measurements in the second row of the table, from Greenstein, Oke, and Shipman (1971).

Greenstein et al. argued that the earlier measurements of the redshift of light from Sirius B are wrong, likely contaminated by light scattered from the far more luminous Sirius A. That is despite the agreement of the measurements using the 100-inch reflector at

Mount Wilson (Adams 1925) and the 36-inch refractor at Mount Hamilton (Moore 1928), because they might have been expected to have been differently affected by scattered light. But Greenstein et al. had measurements when Sirius B had moved further away in the sky from Sirius A, reducing that problem. Their reexamination of the early models and measurements led them to conclude that errors in theory and observation accidentally produced apparent consistency. It happens. It illustrates a hazard that demands repeated independent checks.

Let us pause to note another hazard, this one the potential for confusion in communication between disciplines. Hetherington (1980), an historian, argued that Greenstein et al. (1971) had offered insufficient justification for discarding the earlier redshift measurements, and concluded that

> a strategy of casually dismissing earlier observations for little reason other than their age and disagreement with more recent observations has revolutionary implications for science. It is a question of legitimacy ...

Greenstein et al. (1971) had a legitimate case for dismissing the earlier measurements, beginning with the greater distance of Sirius B from Sirius A at the time of their observations. In their second paper Greenstein, Oke, and Shipman (1985) responded to Hetherington's remarks by making this point and adding more details of their justification. They also emphatically objected to the suggestion that they casually dismissed data, which would have been seriously wrong. I understand the annoyance Greenstein et al. must have felt, and hope Hetherington did not intend to be so offensive.

Yet another aspect of this story is that when Greenstein et al. (1971) presented corrected data for Sirius B there already were the more precise and better controlled tests of the prediction of the gravitational redshift and related effects reviewed in Section 3.4. The interest in white dwarfs had shifted to tests of their structures and the exploration of their formation during the evolution of our galaxy. An example is the Trimble and Greenstein (1972) paper on the redshifts and motions of a considerable number of white dwarfs. Such are the hazards of communication between disciplines, as in the science wars discussed in Section 2.3

It is relevant to the history of tests of general relativity that Popper (1954) concluded that the measured and predicted values of the gravitational redshift of another white dwarf star, 40 Eridani B, agree within the uncertainty of a factor of about 2. Greenstein and Trimble (1972) improved that, reporting measured and predicted redshifts $23 \pm 5$ km s$^{-1}$ and $20 \pm 9$ km s$^{-1}$. But, as noted, this is of interest as a check of the theory of the structures of white dwarf stars, and it is of historical interest concerning how general relativity became empirically established. But the developments in the 1970s were not a significant contribution to the tests of the general theory of relativity.

Examples of early thinking about the tests of the predicted gravitational redshift are to be found in monographs and textbooks on relativity. Tolman's (1934) *Relativity, Thermodynamics, and Cosmology*, Bergmann's (1942) *Introduction to the Theory of Relativity*, Møller's (1952) *The Theory of Relativity*, and McVittie's (1956) *General Relativity and Cosmology*, all reported reasonable consistency of theory and observation of the shift of spectral lines from the sun and Sirius B. Bergmann did not present attributions. Tolman and Møller referred to St. John (1928) and Adams (1925). St. John had reported redshifts of absorption lines in the light from the sun, which we have seen are not readily interpreted, and Adams had reported the test from the white dwarf star Sirius B, which was seriously wrong. McVittie (1956) reported Popper's (1954) conclusion that the theory and observation of the redshift of 40 Eridani B agree, to a factor of 2 or so. McVittie took the radius of Sirius B to be $0.008\,R_\odot$, well below the radius in the first line of Table 3.1. The smaller radius increases the predicted redshift to about 79 km s$^{-1}$. McVittie attributed the radius to Finlay-Freundlich (1954), who referred to the Gamow and Critchfield (1949) estimate of the radius of a model white dwarf with no hydrogen in the interior. That agrees with Kuiper's conclusion. McVittie also reported, in a note added in proof for the book, that

> unpublished examination of this material by G. P. Kuiper indicates that a red-shift of 60–80 km.sec.$^{-1}$ is more probable than these values, which were apparently affected by scattered light from Sirius A. Thus for Sirius B also, there appears to be good agreement between theory and observation.

This would make the theory and observation of the redshift of Sirius B consistent with the Greenstein et al. (1971) results in the second line in Table 3.1.

In the proceedings of the 1955 conference, *Jubilee of Relativity Theory* (this is 50 years for the special theory, 40 for the general theory), Trumpler (1956) cautioned that

> Measures of the solar spectrum at the Mt. Wilson observatory by St. John did suggest an average red-shift of this order [the predicted value]. The discrepancy of various groups of lines and in different parts of the solar surface, however, are considerable, and the confirmation of the theory by the solar observations is not very convincing.

Trumpler reported consistency of theory and observation of the gravitational redshifts of the two white dwarf stars, Sirius B and 40 Eridani B. He missed the clues McVittie had tracked down that the test from Sirius B was wrong.

In the proceedings of the *Chapel Hill Conference* on *The Role of Gravitation in Physics*, Wheeler (1957) reported without attribution the consistency of Popper's (1954) estimates of the predicted and observed redshifts for 40 Eridani B, and McVittie's (1956) positive assessment for Sirius B. Wheeler concluded that "general relativity is not in disagreement with the observations."

A careful review of the redshift tests led Bertotti, Brill, and Krotkov (1962)[7] to report that

> we are not in a position to explain the details of the deviations of the solar red shift from the theoretically predicted value, nor can we determine whether the functional form of the relationship [the prediction that the wavelength shift $\delta\lambda$ is proportional to $\lambda$] is correct. At best, the relativistic effect makes the solar red shift somewhat less puzzling than it would be if we had no theoretical prediction of this kind.... Since the mass of Sirius B nearly equals that of the sun, its radius should be 1/100 solar radius, so that the predicted red shift

7. Krotkov was a member of Dicke's Gravity Research Group. As will be discussed, the group was working to improve the empirical basis for gravity physics. Brill was in Wheeler's relativity theory group at Princeton University, and Bertotti was a visitor at the Plasma Physics Laboratory, Princeton, New Jersey, and prior to that a visitor at the Princeton Institute for Advanced Study. They acknowledged Dicke's help in their assessments of the gravitational redshift and a broad range of other tests of gravity physics.

amounts to 60 km/sec. Observations were made by Adams (1925) and Moore (1928) on the shift of $H_\beta$ and $H_\gamma$ [lines of atomic hydrogen in the atmosphere] in Sirius B under varying conditions of atmospheric scattering. . . . The results range from 9 km/sec to 31 km/sec and seem to be correlated with the intensity of the scattered light of Sirius A (Schröter 1956), indicating that higher values might be obtained if the scattering had been reduced. Thus to date the results on Sirius B are not in disagreement with theory, but inconclusive.

Their estimates of the mass and radius of Sirius B are close to what Gamow and Critchfield (1949) found, and their predicted redshift is close to the Greenstein et al. (1971) prediction and measurement. They do not seem to have been aware of Kuiper's unpublished suggestion that the measurement could be larger, and consistent with the larger prediction, but they had a hint of the larger redshift from Schröter.

We see a tendency to optimism about the gravitational redshift test. I expect this was the combined effect of a modest degree of empirical support with a willingness to believe in an elegant socially constructed theory.

Møller (1957) recognized the possibility of improving the situation by employing the impressive stability of atomic clocks made possible by the stability of atomic energy level transitions, and usually also by the stability of a mode of oscillation of the electromagnetic field in a low-loss cavity. Møller pointed out that the comparison of times kept by atomic clocks on the ground and in a satellite offers the opportunity for a precision test of the predicted gravitational redshift. The vision was realized in the experiment Gravity Probe A (Vessot and Levine 1979) to be discussed in Section 3.4.

### 3.3.3 GRAVITATIONAL DEFLECTION OF LIGHT

The test of Einstein's prediction of the gravitational deflection of light by the mass of the sun was heavily cried up in the early days of general relativity. One still encounters statements that the observations of the effect in the 1919 solar eclipse, an apparent motion of the stars away from the sun as it moved close to the lines of sight to the stars, established general relativity. The situation was more complicated.

The Newtonian and relativity theories predict that the deflection angle of the light from a distant star varies inversely with the distance of the line of sight from the center of the sun at the point of closest approach. In the Newtonian theory the deflection angle at the edge of the sun is 0.87 seconds of arc, or 0.87''. The relativity prediction is twice that, 1.75''. Dyson, Eddington, and Davidson (1920) reported the reduction of observations of shifts of angular positions of stars near the sun during the 1919 solar eclipse. They presented three estimates of the deflection angle extrapolated to the edge of the sun: $1.98 \pm 0.12''$, $1.61 \pm 0.30''$, and $0.93''$. (The uncertainties in the first two measurements are probable errors. If the probability distribution of errors is Gaussian multiply by 1.5 to get the standard deviations.)[8] Dyson et al. gave lowest weight to the last of these values. The nominal mean and standard deviation (weighted in inverse proportion to the variances) of the two acceptable measurements is $1.93 \pm 0.17''$. This is much larger than the Newtonian prediction, which is important. If you trust Gaussian distributions of errors, it is just one standard deviation from the relativity prediction. Dyson et al. concluded that

> the results of the expeditions to Sobral and Principe can leave little doubt that a deflection of light takes place in the neighbourhood of the sun and that it is of the amount demanded by EINSTEIN'S generalised theory of relativity, as attributable to the sun's gravitational field. But the observation is of such interest that it will probably be considered desirable to repeat it at future eclipses.

It was natural to ask whether the deflection of light passing near the sun might be attributed to an increased index of refraction caused by material near the sun. Eddington, Jeans, Lodge et al. (1919) considered the idea, but Jeffreys (1919) concluded that the known phenomena of refraction are "totally incapable

---

8. If the measurement uncertainties really are random then the probability that the mean differs from the true value by more than one standard deviation, either way, is about 30%. The probability that the difference exceeds three standard deviations is about 0.3%.

of accounting for the observed displacements," given physically acceptable amounts of material near the sun.

Among tests from later observations, the 1922 solar eclipse deserves special attention because there were many bright stars near the path of the sun. Lick observations yielded $1.72 \pm 0.11''$ and $1.82 \pm 0.15''$. Campbell and Trumpler (1928) concluded that

> As to the amount of the light deflections and the law according to which these diminish with increasing angular distance from the Sun's center, the observations agree within the limits of accidental observing errors with the prediction of Einstein's Generalized Theory of Relativity, and the latter seems at present to furnish the only satisfactory theoretical basis for our results.

The state of this test received mixed reviews. Tolman (1934) reported the light deflection measurements from the 1922 Lick expedition, which he termed "the most satisfactory data at present." Bergmann (1942) cautioned that

> The predicted deflection amounts to not more than about $1.75''$, and is just outside the limits of experimental error. A good quantitative agreement between the predicted and observed effects cannot be regarded as significant.

Møller's (1952) assessment was that

> the agreement between Einstein's [prediction of the gravitational deflection] and the observations seems to be satisfactory, but, as the effect is just inside the limits of experimental error, one cannot attach too much weight to the quantitative agreement.

Trumpler (1956) listed nine measurements (after rejection of one from the 1919 eclipse, for poor focus, and one from the 1929 eclipse, which has no estimate of the probable error in the measurement). The straight mean is $1.79''$, close to the relativistic prediction of $1.75''$. A useful statistical measure of the goodness of fit of the measurements to the prediction is the sum over the measurements, $\chi^2 = \Sigma (O_i - P)^2 / \sigma_i^2$, where the predicted value of the deflection angle is $P$, and the measured deflections are $O_i$ with the standard deviation $\sigma_i$ in the measurement of $O_i$. On conversion to standard deviations from the tabulated probable errors under the assumption of Gaussian distributions, I get $\chi^2 = 3$. This is well below the expected value, $\chi^2 = 9$, the number of measurements. The small value of $\chi^2$ might be traced to a combination of two effects. One is that the observers made cautious overestimates of measurement uncertainties. This would be unusual; the common experience is that uncertainties are underestimated because they miss the more subtle sources of error. The second effect follows from the necessary discarding of clearly wrong measurements, for example, due to poor focus. The subtle problem with this is the tendency of excess pruning of borderline cases that reduces the scatter among the measurements that are not discarded. This is not a suggestion of wrongdoing; it is a common human tendency.

Trumpler concluded that

> If one considers the various instruments and methods used and the many observers involved, the conclusion seems justified that the observations on the whole confirm the theory.

This is fair enough, though the small scatter of the measurements compared to their stated uncertainties is not reassuring.

Bertotti, Brill, and Krotkov (1962) offered a more cautious conclusion:

> In all of these [eclipse] experiments, a radial shift outward of about the right order of magnitude was observed, and the shift was less for more distant stars. The decrease does not contradict the 1/r law predicted by general relativity, but neither does it give much support to such a dependence on distance. In Fig. 1-14 a straight line would fit the data about as well as a hyperbola. . . . None of the experiments support the Newtonian value $1.75/2 = 0.875$ sec of arc.

Harvey's (1979) reanalysis of digital scans of plates from the 1919 expeditions shows that the information on the photographic plates was capable of a tighter constraint closer to Einstein's prediction than Dyson et al. (1920) had reported. That is not relevant to earlier thinking, of course, but Kennefick (2009), who knew Harvey's reanalysis, put it that Dyson et al. (1920)

> had reasonable grounds for making their central claim that their results were not compatible with Newton's theory but were broadly compatible with Einstein's.

With due attention to the qualifier, "broadly," this is a reasonable assessment of the state of the gravitational deflection test in 1960, when the situation began to improve.

## 3.4 The Empirical Establishment

In the year 1960 the test of the predicted gravitational redshift was in a confused condition. The test of the gravitational deflection of light by the mass of the sun was more reassuring. Bertotti, Brill, and Krotkov (1962) were not very enthusiastic about the evidence that the deflection angle varies in the predicted way with distance from the sun, but if we assume this we might venture that the measured value of the deflection had been determined to be within about 30% of the general relativity prediction. This is greater than the claimed measurement uncertainties, because we must take some account of the suspiciously small scatter of the measurements compared to what would be expected from the stated measurement uncertainties. It means the test of the predicted gravitational deflection offers significant but modest support for general relativity. The one seriously demanding test in 1960 was the prediction, as we may consider it, of the reasonably well-established departure from the Newtonian theory of the motion of the planet Mercury.

The scant nature of the tests in 1960 might have been understood in the physics community, but it is difficult to find published complaints. The prominent exception is Robert Henry Dicke (1957), who announced at the 1957 Chapel Hill Conference on *The Role of Gravitation in Physics* that

> It is unfortunate to note that the situation with respect to the experimental checks of general relativity theory is not much better than it was a few years after the theory was discovered – say in 1920. This is in striking contrast to the situation with respect to quantum theory, where we have literally thousands of experimental checks.... Professor Wheeler has already discussed the three famous checks of general relativity; this is really very flimsy evidence on which to hang a theory.

Dicke reviewed experiments aimed at improving the situation that were planned or in progress in his Gravity Research Group at Princeton University in New Jersey. I joined the group as a new graduate student the following year. The group already was active, and it has remained productive through several generations.

In his doctoral dissertation in the Gravity Research Group, Jim Brault (1962) presented a clean test of the predicted gravitational redshift of light from the sun. He observed a solar sodium absorption line that is strong enough to have formed well above the turbulence near the photosphere, reducing the problem with Doppler shifts. He obtained precision measurements of the line shape near the line center by application of the synchronous detection methods Dicke pioneered and brought to many of the Gravity Research Group experiments. (The technology is reviewed in Peebles 2017.) Brault showed that the line shape is close to symmetric near the line center, and that the line center is independent of position across the face of the sun to about 5%. These are two indications that the measurements have not been corrupted by Doppler shifts or by the presence of other absorption lines at close to the same wavelength. Brault concluded that

> the ratio of the observed red shift to the theoretical value is found to be 1.05 ± .05.

This is one the first two credible tests of the gravitational redshift, from a new experimental technique applied to an old problem, the measurement of the predicted shift of the wavelength of light from the sun.

The other credible test is the laboratory measurement of the gravitational redshift by Robert Pound and Glen Rebka (1960). They concluded that the ratio of frequency shifts, measured to predicted, that is experimental to theoretical, is

$$(\Delta v)_{\text{exp}}/(\Delta v)_{\text{theor}} = +1.05 \pm 0.10. \qquad (3.1)$$

Within the uncertainties of the estimates of measurement errors the two results are essentially the same. But they were obtained in quite different ways, offering the important check for undetected systematic errors.

Pound and Rebka used the Mössbauer effect for their laboratory measurement, and it is worth pausing to see how this elegant effect enabled a test of gravity physics. Consider two identical atomic nuclei, one in an excited energy level, the other in the lower energy ground level. Imagine they are freely moving, initially at the same velocity. When the nucleus in the excited energy level decays to the ground level it emits a $\gamma$-ray photon. The momentum of this decay photon is balanced by the recoil momentum of the nucleus. Since the kinetic energy of this recoil is taken from the decay energy of the nucleus, the $\gamma$-ray photon has a little less than the full difference of energy between the excited and ground levels. This means the photon is not quite energetic enough to be absorbed by the second nucleus, which would promote it from the ground to the excited level. But Mössbauer (1958) demonstrated, by experiment and the prediction of quantum theory, that if the two atomic nuclei are bound in a crystal then the recoil momentum can be taken up by the momentum of the whole crystal or a large part of it. Because the mass of the crystal is enormous compared to an atomic nucleus, the recoil kinetic energy is negligible in this case; in effect the decay is recoilless. This means the decay $\gamma$-ray photon has the full energy difference of excited and ground levels, and can be absorbed by another atomic nucleus in the ground level in the crystal, provided that recoil at absorption also is taken up by the crystal. In this case the decay photon could be absorbed and the second nucleus promoted to the excited level. This is the Mössbauer effect.

Now imagine two of these crystals are at different distances above the ground. When a decay $\gamma$-ray photon from the lower crystal reaches the higher one, the predicted gravitational redshift would have caused the photon to lose energy. (Recall that quantum physics says a photon with frequency $\nu$ has energy $h\nu$, where $h$ is Planck's constant. Lower energy means lower frequency, which is larger wavelength: a redshift. We are combining quantum and relativity physics.) The loss of energy reduces the chance of recoilless absorption of the photon in the higher crystal. But if the higher crystal is moving toward the lower one at just the right speed, so the Doppler shift compensates for the gravitational redshift, then the photons have just the right energy for recoilless absorption. So measure the speed for maximum absorption.

Pound and Rebka (1960) obtained a 10% detection of the gravitational frequency shift of $\gamma$-rays rising or falling 74 feet through helium in a tower at Harvard University. The precision is comparable to Brault's, but the experimental situation is very different.

As I said, this is a valuable check for possible systematic errors in both measurements, an obsession for those interested in assessing critical measurements.

We should pause to consider what the science community would have concluded if the Pound and Rebka measurement had not agreed with the relativity prediction. The Mössbauer effect follows from quantum physics in situations where this theory had been well tested. The effect soon was applied to measurements of properties of condensed matter, with no surprises. The observed temperature shift of the Mössbauer resonance is the second-order Doppler shift; the Pound and Rebka redshift measurement used the first-order Doppler shift. In short, the application of the Mössbauer effect to test the gravitational redshift rested on firm experimental and theoretical ground. If there had been a disagreement between relativity theory and the measurement, and if there had not been Brault's independent check, the community would have blamed the far less well tested relativistic theory. If the Brault measurement agreed with the theory and the Pound and Rebka measurement disagreed, it would have been a fascinating crisis.[9] As it happened the two quite different methods of measurement of the gravitational redshift agreed with each other and with the prediction of the theory. This made a good case that the detections of the relativity prediction could be trusted, within measurement uncertainties.

The laboratory measurement could be improved. Pound and Snider (1964) brought the laboratory precision to 1%, a considerable advance. The experiment Gravity Probe A, conducted along the lines Møller (1957) envisioned, compared rates of atomic hydrogen maser clocks on the ground and carried in a rocket 10,000 meters above the ground. Microwave links up and down compared the times kept by the two clocks. The measurements in the first report by Vessot and Levine (1979) agree with the general relativity prediction to a few parts in $10^4$. Vessot, Levine, Mattison et al. (1980)

---

9. The article on Scientific Realism in the Stanford Encyclopedia of Philosophy (Chakravartty 2017) discusses in Section 3.1 the possibility of such a crisis that calls for an adjustment of some part of a theory. It has not happened yet in the general theory of relativity. Particularly interesting cases where it did happen are the adjustments of the classical theories to include relativity and quantum physics. These adjustments became credible in the usual way, by passing abundant demanding tests of predictions.

improved that by a factor of 10. The precision testing of general
relativity was well started.

Irwin Shapiro (1964) proposed a "Fourth Test of General Relativity" by measurements of the times elapsed from emission to
reception of radar pulses reflected from the planets Venus and Mercury. When the line of sight passes close to the sun, general relativity
predicts an increase in the time elapsed. Shapiro, Pettengill, Ash
et al. (1968) reported clear detection of the Earth-Mercury time
delay near the sun, to about 20% uncertainty, consistent with the
relativistic prediction. A later application of this test measured
the round-trip time of radio signals transmitted from Earth to
transponders on or orbiting Mars that returned the signal to Earth.
The increased time delay when the line of sight passed close to the
sun agrees with general relativity to a few parts in $10^3$ (Reasenberg,
Shapiro, MacNeil et al. 1979). When the line of sight to the Cassini
spacecraft on its way to Saturn passed close to the sun, the radio frequency shift resulting from the Shapiro time delay again was found
to agree with general relativity. It is pleasing that this analysis, by
Bertotti, Iess, and Tortora (2003), yields a clear demonstration of
the expected variation of the frequency shift effect with distance
of closest passage by the sun (in their fig. 2). Four decades earlier,
when Bruno Bertotti was working with members of Dicke's Gravity Research Group and Wheeler's relativity theory group, Bertotti,
Brill, and Krotkov (1962) had pointed to the absence of a good
demonstration of this scaling with distance for the gravitational
deflection of light.

Radio interferometry allowed a much better test of the gravitational deflection of electromagnetic radiation by the mass of the
sun. The first applications to the angular displacement of the quasar
3C279 as the line of sight passed near the sun reported that the
measured deflections, translated to the edge of the sun, are $1.77 \pm$
$0.21$ arc sec (Seielstad, Sramek, and Weiler 1970) and $1.82^{+0.26}_{-0.18}$
arc sec (Muhleman, Ekers, and Fomalont 1970). The results are
essentially the same and consistent with the relativistic prediction,
1.75 arc sec, to about 15%. Clifford Will's (2014) survey of the tests
of relativity reports very long baseline radio interferometer measurements that agree with the general relativity prediction to a few
parts in $10^4$.

These are examples of the revolutionary advances in empirical gravity physics that began in 1960. They grew out of old and new ideas, and they relied on advances in technology: Brault's use of precision synchronous detection, Pound and Rebka's use of the Mössbauer effect, Shapiro's use of radar powerful enough to reach other planets, Vessot and Levine's use of atomic clocks, and the list goes on.

## 3.5   The Lessons

From 1915 to 1960 the anomaly in the Newtonian theory of the orbit of the planet Mercury was more securely measured than the gravitational deflection of light, which was more secure than the failed attempts to find a meaningful test for the gravitational redshift. Yet the anomaly in the orbit received less attention than the gravitational deflection, and even the gravitational redshift. From the present perspective this seems irrational, but we all tend to be irrational at times.

Following Latour and Woolgar's (1979) original terminology, we may say that Einstein's general theory of relativity in 1960 was a social construction. In Dawid's (2013) terminology, general relativity was non-empirically confirmed. Despite the modest empirical support the considerations in Section 3.1 made general relativity a standard and accepted part of theoretical physics. This seems irrational too, but it was difficult to resist the non-empirical consideration that electromagnetism is the simplest relativistic vector theory (that is, $A_\mu$) for conserved charge, and general relativity is the simplest relativistic tensor theory (that is, $g_{\mu\nu}$) for local conservation of energy and momentum. Electromagnetism passed abundant tests. It made sense to pay special attention to general relativity.

I have heard it said that the general theory of relativity was there, waiting to be discovered. Ernst Mach surely would have rejected this as empty metaphysics, but we must consider that the theory Einstein and Hilbert considered so elegant a century ago, and the scientific community accepted on non-empirical grounds a half century ago, now passes tests of predictions on scales ranging from the laboratory to the solar system and on to the observable universe (as to be reviewed in Sec. 6.10). It is remarkable that pure thought sometimes points us in the right direction; here is an example. But of course it is not at all surprising that pure thought more often leads us astray. Einstein did not set us an edifying example by discovering general relativity by close to pure thought.

We have been lucky that it has been possible to demonstrate that the general theory of relativity passes a considerable variety of precision tests applied on a vast range of length scales. It is obvious, but must be said, that in light of this evidence dismissing general relativity as a social construction would require the postulates of an absurd variety of the Putnam miracles discussed in Section 1.3.

I conclude that Einstein's theory has all the appearances of a good approximation to objective reality. The theory is incomplete—the usual situation—but the case for it is compelling.

CHAPTER FOUR

# Einstein's Cosmological Principle

Disagreements between theory and practice may be resolved either way. The history of thinking about Einstein's cosmological principle, the assumption that the universe is the same everywhere on average, is an example. When Einstein introduced arguments for this picture he may not have known that it was quite contrary to the observations. After publication of the proposal, de Sitter told Einstein about the contrary evidence, but Einstein, and with him the community by and large, continued to accept the assumption. It was a social construction. The early observations were consistent with the idea that matter is organized in a hierarchy of clusters within clusters of clusters and on up. But by the time some started to take this picture of a hierarchy of clusters seriously, the improving observations were pointing to Einstein's idea of large-scale homogeneity. The evidence for Einstein's picture looks compelling now.

## 4.1  Einstein's Homogeneous Static Universe

Ernst Mach's (1883, 1902) critical examination of the science of mechanics includes his argument that velocity and acceleration surely have no intrinsic, objective, significance; they are meaningful only relative to the behavior of the rest of matter. The concept became known as Mach's principle. It has been influential to Einstein and others, though Einstein later regretted it, and the possible physical significance of Mach's argument still is debated.

In the years around 1920 Einstein's (1917) considerations of Mach's argument led him to propose that a logically arranged universe would be the same everywhere, apart from local fluctuations such as stars and planets and people. In this close to uniform, spatially homogeneous, universe there would be no edge and no preferred center. To see the reason for this thinking, suppose instead that, at great distance from an island of matter, the universe is empty and has the flat spacetime of special relativity. As one says, suppose spacetime is asymptotically flat. Einstein remarked that

> If only one single point of mass were present, according to this view [asymptotically flat spacetime], it would possess inertia, and in fact an inertia almost as great as when it is surrounded by the other masses of the actual universe.

To put it more broadly, general relativity allows a universe in which there is a single massive body, a galaxy, outside of which spacetime is empty and asymptotically flat. This galaxy could rotate, with all the usual effects of rotation, but it would be rotation relative to empty spacetime. In Einstein's 1921 Princeton lectures on *The Meaning of Relativity* (published in Einstein 1922a) he concluded that, if the universe were constructed this way, then

> Mach was wholly wrong in his thought that inertia, as well as gravitation, depends upon a kind of mutual action between bodies.

In the general theory of relativity, inertial motion in an asymptotically flat spacetime is determined by two things: the nature of the material contents of the universe and the boundary condition that at great distance from all the matter spacetime is flat. We have an indication of Einstein's thinking about boundary conditions— what the universe is like far away—in a letter to Michele Besso[1] dated 22 April 1916:

---

1. The quotations from letters here and in the following considerations are from the Collected Papers of Albert Einstein Vol. 8, Document numbers 219, 272, 273, 293, and 321, in the order discussed here.

> In gravitation I'm now looking for the boundary conditions at infinity; it certainly is interesting to consider to what extent a *finite* world exists, that is, a world of naturally measured finite extension in which all inertia is truly relative.

So Einstein was considering whether his new theory might do away with boundary conditions. Might general relativity allow a universe that is consistent with Mach's argument that inertial motion and the departures from inertial motion that we term accelerations are defined solely by the contents of the universe? At about this time Einstein was considering the thought that spacetime ends where the components of the metric tensor $g_{\mu\nu}$ approach infinity or zero, so length and time intervals are meaningless. This would be the degenerate condition of spacetime at what Einstein supposed are distant hypothetical masses.

Willem de Sitter, in the Netherlands, did not like the idea. In a letter to Einstein dated 1 November 1916 he wrote that

> it is very hard for me to believe in the distant masses. I would prefer having no explanation for inertia to this one.

Similar sentiments are expressed in de Sitter's (1916) paper, *On Einstein's Theory of Gravitation, and its Astronomical Consequences.*

In a letter dated 4 November 1916 Einstein agreed with de Sitter, abandoned the hypothetical masses, and added that

> you must not scold me for being curious enough still to ask: Can I imagine a universe or the universe in such a way that inertia stems entirely from the masses and not at all from the boundary conditions?

On 2 February 1917 Einstein reported to de Sitter that

> Presently I am writing a paper on the boundary conditions in gravitation theory. I have completely abandoned my idea on the degeneration of the $g_{\mu\nu}$'s, which you rightly disputed. I am curious to see what you will say about the rather outlandish conception I have now set my sights on.

The paper mentioned in this letter, Einstein (1917), was received at the journal six days later, 8 February. In it Einstein proposed to replace boundary conditions by the assumption that the universe is the same everywhere, apart from minor fluctuations such as stars and planets: no boundary, no boundary conditions.

The idea of a near uniform, homogeneous, distribution of matter everywhere is elegant. One might expect that this near homogeneous distribution of matter naturally would produce a homogeneous and isotropic spacetime, meaning that rotation would be relative to the spacetime set by the matter, as Mach would have it. (To be technical, spacetime is not uniquely defined by the presence of a homogeneous mass distribution: the mass can shear and rotate as well as expand. But that entered considerations later.) Maybe this is what Einstein was thinking when he wrote the 4 November letter to de Sitter, and surely what he reported finding in the 2 February 1917 letter.

In a paper marked "Communicated in the meeting of March 31" to the Royal Dutch Academy of Sciences, de Sitter (1917) discussed Einstein's new idea about the universe, and wrote that

> The idea to make the four-dimensional world spherical in order to avoid the necessity of assigning boundary-conditions, was suggested several months ago by Prof. EHRENFEST, in a conversation with the writer. It was, however, at that time not further developed.

Einstein and Ehrenfest were exchanging letters. I do not know whether they exchanged views on the idea of a homogeneous universe.

In his paper on the starting idea for modern cosmology, homogeneity, Einstein (1917) presented another argument for his idea. He took it without comment that the universe is unchanging. Stars move around, but in his picture this is in a spacetime that is on average independent of time. He pointed out that an isolated gravitationally bound star cluster would evolve by gravitational interactions among the stars that would give some escape velocity. This evaporation means the cluster could not last forever. Einstein took this to mean that there must be stars everywhere, in a homogeneous universe, with no clusters and so no evaporation. But it was already clear that this argument is wrong, and the idea of homogeneity is seriously challenged, on several grounds, as follows.

Einstein could have consulted astronomers about his idea of a homogeneous universe. He was in Berlin at the time, 1916 and

1917, during the Great War, but the Netherlands was neutral. That allowed him to exchange mail with Willem de Sitter and others at Leiden Observatory and University. Correspondence also was possible between de Sitter and Eddington, professor of astronomy at the University of Cambridge, England. Eddington could, perhaps did, tell Einstein about his book, *Stellar Movements and the Structure of the Universe* (Eddington 1914). The universe in this book is our Milky Way galaxy of stars. Astronomers knew from star counts that our galaxy has edges and a central concentration of stars, not at all like Einstein's vision of the universe. Eddington's book has astronomical photographs of some spiral nebulae. He suggested that they are other galaxies of stars, as they prove to be. But astronomers knew that the spiral nebulae, an early term for galaxies, are distributed across the sky in a decidedly clumpy way.

After publication of Einstein's (1917) proposal that the universe is homogeneous, de Sitter wrote to Einstein (in a letter dated 1 April 1917) to protest that

> I believe that it is probably certain that even the Milky Way is not a stable system. Is the entire universe then likely to be stable? The distribution of matter in the universe is extremely *inhomogeneous* [I mean of the stars, not your "worldmatter"], and it cannot be substituted, even in rough approximation, by a distribution of constant density. The assumption you tacitly make that the mean stellar density is the same throughout the universe [naturally, for vast spaces of, e.g., (100,000 light-years)] has no justification whatsoever, and all our observations speak against it.

The square brackets in this quotation are de Sitter's, and de Sitter's remarks remain sensible.

Einstein (1917) understood that a gravitationally bound concentration of stars such as the Milky Way cannot last forever; he mentioned it. Possibilities are that in February, when he published his arguments for homogeneity, Einstein did not know we are in a concentration of stars, or perhaps he knew but supposed the Milky Way is not gravitationally bound. By an "extremely *inhomogeneous*" distribution of matter, de Sitter may have had in mind the bounded and far from spherical distribution of stars in the Milky Way, or maybe the markedly inhomogeneous distribution of the spiral nebulae that some argued are other galaxies of stars. In any case, given Einstein's independence of mind, it would not be surprising if he ignored whatever astronomical evidence he knew in favor of his elegant idea.

It also would not be surprising if Einstein did not know James Jeans' (1902) demonstration that, in Newtonian physics, a static unbounded homogenous mass distribution is exponentially unstable to the growth of departures from homogeneity. This demonstration is sometimes called the Jeans swindle, because it is difficult to know what to make of an infinite homogeneous mass distribution in Newtonian gravity, but Jeans did the sensible if mathematically awkward thing. In general relativity, Einstein's picture of an unbounded near homogeneous and static universe is a well-defined situation, and Jeans' analysis readily carries over. This is because the Newtonian gravity theory Jeans used is the correct limiting case of general relativity for a small patch of this near homogeneous universe. Einstein learned only later that his relativistic static near homogeneous universe is Jeans unstable. This is another reason why it cannot last forever. (Lemaître 1927 showed that Einstein's static universe is unstable against a homogeneous perturbation from equilibrium, and Lemaître 1931 demonstrated the instability for a spherically symmetric perturbation from a homogeneous distribution of pressureless matter.)

Einstein also failed to recognize that, if energy were conserved, the lifetimes of stars would be limited by the energy available, which cannot exceed the annihilation energy of the stellar masses. Conservation of energy would mean the universe of stars could not last forever. If energy were not conserved, and stars had been shining forever, then in a static spatially homogeneous universe there would have to be some provision to dispose of the indefinitely long accumulation of starlight. This is Olbers' paradox. There is no paradox of this sort in an expanding homogeneous universe, because this universe has a limited time of expansion since early conditions in which stars could not have existed. (This time is now put at a little more than ten thousand million years, about the lifetime of a star with the mass of the sun.)

Janssen (2014) presents still more considerations of Einstein's problematic picture of a universe that is homogeneous and static.

These problems could have been recognized as a hint to a ready solution, that the universe is expanding. I do not know whether anyone thought of this before more direct evidence of expansion was noticed. But let all this not obscure the fact that Einstein was right about homogeneity. The question remains, was he right for the right reason?

## 4.2 Evidence of Homogeneity

Edward Arthur Milne (1933) termed Einstein's assumption that the universe is the same everywhere the "extended principle of relativity" and "Einstein's cosmological principle." The latter became standard. Milne also added theoretical weight to the argument for homogeneity, despite the apparently conflicting astronomical evidence. Milne demonstrated that, if the universe is expanding in a homogeneous and isotropic way, the speed at which a galaxy moves away from us is proportional to its distance. Lemaître (1927) showed that this relation between galaxy distance and its velocity of recession follows from homogeneity in general relativity. Milne proved it from homogeneity for any theory in which distance and relative velocity have the usual meanings. The effect was observed. Hubble and Humason (1931) found that the proportionality predicted by Lemaître and Milne is a good approximation for the most luminous galaxies observed out to distances so great that the recession speed is six percent of the speed of light. This was a considerable advance over Hubble's (1929) evidence in his first argument for the relation. Hubble and Humason presented an impressively deep probe into the expanding universe.

The relation between recession speed and distance has been variously known as the law of the general recession of the galaxies, Hubble's law, and now the Hubble-Lemaître law. (The relation is $v = H_0 r$, where $v$ is the speed of recession of the galaxy derived from the observed redshift interpreted as a Doppler shift, $H_0$ is the constant of proportionality, Hubble's constant, and $r$ is the distance to the galaxy.[2])

2. To be more careful, we should note that the relation $v = H_0 r$ applies after correction for the motions caused by the pull of the gravity of local mass concentrations. The distance

Although Hubble's law is consistent with homogeneity, it is not required. In an earlier paper, Milne (1932) pointed out that the same relation follows by velocity sorting. Imagine a great explosion gave the galaxies in an original concentration randomly directed velocities with a large range of speeds. And ignore gravity. Then faster moving galaxies move further away, approaching the situation in which the distance of a galaxy is proportional to its speed of recession, which is Hubble's law.

Fritz Zwicky (1929) proposed another thought, that light may suffer a "drag" as it moves though space, lowering the energy and hence the frequency. The further light traveled, the greater the integrated drag, the lower the frequency, and the larger the wavelength shift. It is Hubble's law again, with no need for homogeneity. These were early times in cosmology.

Hubble presented two empirical tests for homogeneity. One of them uses the prediction that, if the spatial distribution of galaxies is homogeneous on average, then the count of galaxies brighter in the sky than $f$ varies as $N(>f) \propto f^{-3/2}$. Astronomers had used this relation[3] to show that the distribution of stars in our Milky Way galaxy is decidedly not homogeneous: our galaxy has edges. Hubble's (1936) galaxy counts increase with decreasing brightness

---

$r$ between two galaxies is physical, in an idealized way. It is the sum of the lengths found by a line of observers between the two galaxies, each equipped with a tape measure, and each of whom agrees to measure the distance to the next in line at a given cosmic time. The speed is the rate of change of this physical distance according to a physical clock. At distances large enough that relativistic effects are important, this idealized measurement of the instantaneous distancer between the galaxies is increasing more rapidly than the speed of light. This is consistent with relativity; no single observer sees this rate of separation. An observer's measure of the expansion is the redshift-magnitude, or $z$-$m$, relation. The redshift, $z$, a measure of the velocity of recession, is defined in footnote 12 on page 169. The apparent magnitude, $m$, a measure of the distance, is defined in footnote 7 on page 99.

3. A measure of the intensity of the starlight from a galaxy is the rate $f$ of reception of starlight energy per unit area. This is the energy flux density. To see how the count of galaxies brighter than $f$ varies with $f$ suppose first all galaxies have the same luminosity, $L$. If a galaxy is at distance $r$ then we observe starlight energy flux density $f = L/(4\pi r^2)$. Given $L$, if the flux density is $f$ then the distance is $r \propto f^{-1/2}$. If these galaxies are uniformly distributed through space then the number counted brighter than $f$ is the product of the number density with the volume of space to distance $r$, or $N(>f) \propto r^3 \propto f^{-3/2}$. Since this is the same power law form for all galaxy luminosities, it must apply to the sum over all galaxy luminosities. This assumes the Euclidian geometry of flat space, which is an excellent approximation at the distances Hubble could explore.

in the sky, that is, decreasing $f$, meaning increasing distance on average, somewhat more slowly than expected for homogeneity. That could be an indication of lower number densities of galaxies at greater distances, but later evidence is that it is the result of a systematic error in Hubble's estimates of the relative brightnesses of galaxies in the sky. This is difficult to get right. But Hubble was counting galaxies out to distances so great that the redshifts indicate speeds of recession approaching ten percent of the speed of light. These observations of very distant galaxies did not suggest Hubble was reaching an edge to the universe of galaxies.

Hubble's second test compares counts of galaxies brighter than a chosen value, $f$, in different directions in the sky. This is not easy but it is less hazardous than calibrating ratios of values of $f$ for galaxies with different brightnesses in the sky. The spatial distribution of galaxies was known to be quite clumpy on the relatively small scales that were observed when Einstein and de Sitter were first thinking about this. If the distribution approached homogeneity in the average over larger distance scales, then the counts of galaxies across the sky would be found to be close to uniform when the galaxies are counted to distances large enough to average out the clumps. Hubble (1934) reported that this is what he saw.

> On the grand scale, however, the tendency to cluster averages out. The counts with large reflectors conform rather closely with the theory of sampling for a homogeneous population. Statistically uniform distribution of nebulae appears to be a general characteristic of the observable region as a whole.

In the 1950s the thinking about the case for Einstein's uniform mass distribution from Hubble's observations is illustrated by the comments in two influential books on relativity and cosmology. Though I have mentioned that there is little consideration of phenomenology in the Landau and Lifshitz series of books on theoretical physics, my edition of *The Classical Theory of Fields* offers a sensible cautionary footnote:

> Although the astronomical data available at the present time give a basis for the assumption of uniformity of this density, this assumption can of necessity have only an approximate character, and it remains

an open question whether this situation will not be changed even qualitatively as new data are obtained, and to what extent even the fundamental properties of the solutions of the equations of gravitation thus obtained [from the assumption of homogeneity] agree with actuality.

In the book *Cosmology*, Bondi (1952, pp. 14 and 15) remarks on the clumpy distribution of the galaxies, but notes that Hubble and Humason's (1931) redshift-distance measurements are consistent with homogeneous expansion, as Milne (1933) had shown. And Bondi points out that Hubble's (1936) counts of galaxies to the most distant that could be detected shows no serious evidence that the observations are approaching an edge to the universe of galaxies. Bondi could have added the near isotropy of Hubble's (1934) deep galaxy counts.

Beginning in the 1950s the case for large-scale homogeneity was being improved by indirect evidence from observations at wavelengths longer and shorter than optical. Some galaxies are strong sources of radio radiation (now considered to be the result of violent compression of matter falling into the massive black hole at the center of the galaxy). These radio galaxies are detectable at great distances. The *Second Cambridge Catalog of Radio Sources*, usually known as 2C, includes about 500 sources (detected at frequency 81.5 MHz) with reasonably well determined angular positions. They are quite close to uniformly distributed across the sky, as seen in Figure 3 in Shakeshaft, Ryle, Baldwin et al. (1955). This is the same as the impression Hubble (1934) had from the near uniform optical counts of distant galaxies across the sky.

In the 1950s only a few of these radio sources were known to be galaxies. It added uncertainty to the test: are these radio sources really distant galaxies? Improved measurements of source angular positions, in the revised third radio source catalog 3CR (Bennett 1962), with improved optical detector efficiency, allowed Smith, Spinrad, and Smith (1976) to report that 137 of the 328 3CR sources are distant galaxies, and 50 more are distant quasars. The latter are now termed active galactic nuclei: quasars centered in galaxies. The close to uniform distribution of these radio galaxies across the sky agrees with the idea that their space distribution has averaged out to uniform at the typically large distances of these galaxies.

There are alternative interpretations. Maybe the universe is inhomogeneous but spherically symmetric about our position. Maybe radio-loud galaxies are uniformly distributed while ordinary galaxies are not. But both ideas seem unlikely.

Another line of evidence in the 1960s came from the discoveries that we are in a sea of radiation, observed at X-ray and microwave wavelengths. At both wavelengths the radiation is quite close to uniformly distributed across the sky. Again, this suggests the large-scale spatial uniformity of mass, the cosmological principle.

More direct and quantitative measures of the spatial distribution of the galaxies grew out of catalogs of the angular positions of galaxies with rough estimates of their distances. Projects that began in the 1950s led to Fritz Zwicky's *Catalogue of Galaxies and of Clusters of Galaxies* (Zwicky, Herzog, Wild et al. 1961–1968), the deeper *Lick Galaxy Catalog* compiled by Donald Shane and Carl Wirtanen (1967), and the still deeper map of the richest clusters of galaxies identified by George Abell (1958). At the 1958 Solvay Conference on *La structure et l'évolution de l'univers,* Jan Oort (1958) reviewed the evidence from these surveys then in progress. Oort concluded that

> It appears, then, that on this scale (diameters of volumes about 350 million pc)[4] there remains little, if any, real variation in mean density. As this volume is still small compared to the dimensions of the universe [the Hubble length], we may conclude that all available evidence supports the concept of a universe which is homogeneous on a large scale.... It should be mentioned that a similar conclusion had already been obtained by Hubble 25 years ago [Hubble 1934] from counts of individual nebulae down to about the same distance. The measurements of the velocities of recession of distant nebulae point to the same conclusion.

In the 1970s members of the Princeton Gravity Research Group were reducing the data[5] from the surveys that were in progress

---

4. This is 350 Mpc, where one megaparsec, 1 Mpc, is $3.086 \times 10^{24}$ cm, about 10 million light years. The Hubble length, at which the redshift-distance relation extrapolates to the speed of light, is about 4000 Mpc.

5. The data in the particularly important Lick Galaxy Catalog were compiled by Shane and Wirtanen (1967), reduced by Seldner, Siebers, Groth, and Peebles (1977), and analyzed along with the other two galaxy catalogs by Groth and Peebles (1977).

when Oort mentioned them, along with the still deeper Jagellonian field (Rudnicki, Dworak, Flin et al. 1973), to statistical
measures of the galaxy distribution in space and of the galaxy
radial velocities relative to the smooth Hubble flow. The galaxy
position correlation functions turn Hubble's (1934) argument for
large-scale uniformity into a quantitative measure. The galaxy
distribution is found to be distinctly clumpy on scales less than
30 million light years, but the distribution averages out to uniform when smoothed over larger scales.[6] In the jargon, the galaxy
space distribution is well approximated as a stationary random
process.

Similar statistical considerations relate the gravitational effect
of departures from an exactly homogeneous mass distribution to
the departure of the sea of microwave radiation from an exactly
isotropic distribution across the sky. Detection of this anisotropy
indicates the mass distribution varies across the universe we can
observe by about one part in one hundred thousand ($\delta\rho/\rho \sim 10^{-5}$).
This assumes Einstein's general theory of relativity, of course, but
the many cross-checks of measures such as this offer a case for this
assumption that is compelling for the physical cosmology community, and I would say an excellent bet for the rest of us. The situation
is discussed in Chapter 6.

## 4.3 The Fractal Universe

Einstein (1917) arrived at the picture of a homogeneous universe
by pure thought. De Sitter, who he respected, told him that this was
not what the observations suggested, but Einstein's picture proves
to pass later demanding tests. Gérard de Vaucouleurs, an observational astronomer, and Benoît Mandelbrot, a mathematician, were
influential sources for a growing interest in another idea, that the
galaxies are in clusters that are in superclusters of clusters and so
on up, maybe in a clustering hierarchy that extends to indefinitely
large scales and indefinitely low mass density in the average over

6. To be quantitative: a sphere of radius 10 Mpc, placed at random, contains counts of
galaxies that scatter about the mean, $N$, by the root-mean-square value $\delta N = N$. On scales
smaller than this radius the galaxy distribution is decidedly clumpy. On much larger scales
the distribution averages out to uniformity, as Oort reported in 1958 and Hubble in 1934.

large enough length scales. The early evidence suggested this. In the 1970s Mandelbrot presented forceful arguments for the elegance of the clustering hierarchy picture, and de Vaucouleurs argued for its consistency with the observations of how matter is distributed. To be considered here are these arguments and the evidence they are wrong.

A century ago Charlier (1922) presented a careful analysis of the clustering hierarchy picture and the evidence apparently pointing to it. He took the spiral nebulae to be galaxies similar to our Milky Way, as did Eddington in 1914, though the idea was not yet generally accepted. It would mean that stars are in clusters, the galaxies; the galaxies are in the clusters that were observed then; and maybe these clusters of galaxies are in superclusters; and so on up, a clustering hirearchy.

Charlier led the creation of a map of positions across the sky of 11,475 extragalactic nebulae. He removed star clusters and other concentrations of starlight within the Milky Way, though some likely remain. The map shows very few nebulae close to the plane of the Milky Way. Charlier proposed that this is caused by

> dark matter lying in the plane of the Milky Way and hiding the nebulae, or a great part of the nebulae, in this direction. The many nebulae photographed at Mount Wilson and at other observatories having a dark band along the thin edge of the nebula give a good support for such a suggestion.

This zone of avoidance in the Milky Way is indeed the result of obscuration of starlight by a form of dark matter, interstellar dust that is close to the plane of the Milky Way. As Charlier indicated, the same effect is seen in other spiral galaxies.

Charlier remarked that a

> remarkable property of the image [the map of the distribution of nebulae, or galaxies] is that the nebulae seem to be piled up in clouds (as also the stars in the Milky Way). Such a clouding of the nebulae may be a real phenomenon, but it may also be an accidental effect either caused by dark matter in the space or declared by condensation of the observations on singular points of the sky.

This assessment is cautious, but the clumpy distribution of the nebulae, or galaxies, in Charlier's map is real, and it certainly suggests a clustering hierarchy.

Harlow Shapley and Adelaide Ames (1932) made a map of the distribution across the sky of the 1,025 galaxies brighter in the sky than a fixed value (brighter than apparent photographic magnitude[7] $m = 13$). They too remarked on the clumpy distribution, including the great difference in numbers of galaxies above and below the plane of the Milky Way. Again, the map suggests a clustering hierarchy.

Gérard de Vaucouleurs, an accomplished observer, knew that Charlier's dark matter in the Milky Way and other galaxies is interstellar dust, and that obscuration by this dust is responsible for the scarcity of observed galaxies in directions close to the plane of the Milky Way, the zone of avoidance. De Vaucouleurs (1970) also knew that the clouds of galaxies seen above and below the plane are real, and he pointed out that these observations are most directly and simply interpreted as part of a clustering hierarchy. He took note of Hubble's (1934) deep counts of galaxies in samples across a good part of the sky, pointing out that the counts vary from sample to sample far more than expected if the galaxies were placed at random (in a stationary random Poisson process). The positions of the galaxies are indeed strongly correlated on relatively small scales, consistent with this scatter from sample to sample. But de Vaucouleurs did not mention the evidence from Hubble's (1934) observations, and the angular distribution of radio galaxies, that the clustering averages out on large scales. The position de Vaucouleurs took in 1970 was defensible, but already there were the challenges that Oort (1958) reviewed in his assessment of the situation.

Benoît Mandelbrot recognized the common occurrence of patterns that bring to mind clustering hierarchies; he named them fractals. An example in his book in several editions, *Les objets*

---

7. A measure of the brightness in the sky of a star or galaxy is the rate, $f$, of starlight energy received per unit area. The astronomers' measures of the apparent magnitude, $m$, given the energy flux density, $f$, and the absolute magnitude, $M$, given the luminosity, $L$, are

$$m = -2.5 \log_{10} f + \text{a constant, and } M = -2.5 \log_{10} L + \text{another constant.}$$

We need not bother with the two constants.

*fractals* (Mandelbrot 1975, 1989), is the length of the coastline of Brittany. It behaves as a fractal, because the length depends on the spatial resolution at which it is measured: finer resolution picks up more length in bends and turns along the landscape and increases the length. Mandelbrot's stroke of genius in recognizing the great range of applications of the concept of fractals and his energetic promotion of the idea forced broad appreciation.

Perhaps it was inevitable that Mandelbrot should consider the idea that the galaxy distribution is fractal, and it certainly helped that he could quote supporting evidence from a respected observer, de Vaucouleurs. The combination was effective; the fractal universe attracted considerable attention. One might be tempted to say that reality missed the chance to present us with something more interesting than homogeneity: a fractal universe. The evidence leading up to the tests reviewed in Section 6.10 is clear, however. The universe we can observe is on average the same everywhere, a good approximation to a stationary random process.

## 4.4   Lessons

We see the power of ideas in the ways the cosmological principle and the fractal universe were welcomed or rejected in different quarters. Mandelbrot had the genius to recognize the widespread relevance of fractals. If he had applied the vigor of his arguments for a fractal universe two decades earlier it would have been a more productive stimulus to research. As it is, when the idea was promoted it was already challenged by the isotropy of the radiation backgrounds— radio, microwave, and X-ray—and by the observations of the galaxy distribution that Oort (1958) reviewed. Despite this the fractal universe attracted considerable attention, because the idea is interesting, and fractals have many other useful applications.

We also see the power of ideas in the respectful attention that de Sitter and Eddington gave to Einstein's picture. They both knew the astronomical evidence that does not bring to mind Einstein's uniform universe, however elegant the concept, and we have seen that de Sitter was frank in telling Einstein about it. But de Sitter and Eddington soon were writing papers about the properties of Einstein's homogeneous universe, and the cosmology community largely followed in accepting homogeneity without complaint. An example is Bondi's (1952) opinion noted in Section 4.2.

The idea of a universe that is the same everywhere was not original with Einstein, but it had been confused by the difficulty of understanding how to apply Newtonian physics to the unbounded mass of an unbounded universe. One might have said that a homogeneous mass distribution without bound is theoretically excluded, except that Jeans found a way to work around the problem. It is notable, and elegant, that a universe that is uniform without bound is described in a mathematically clear way in Einstein's general theory of relativity. This may have influenced some of the early cosmologists, and maybe some were attracted by the convenience of an analytic solution to Einstein's field equation for a uniform universe, but in any case Einstein's idea of homogeneity was deeply influential. The idea was a social construction, and I count it as a Merton singleton, not because the idea was new, but because the idea at last logically fit a theory.

In the 1930s Hubble displayed just the right intuition in the exploration of what he termed the Realm of the Nebulae. He produced empirical support for the cosmological principle from the isotropy of deep galaxy counts, the scaling of the counts with brightness in the sky, and the redshift-distance relation, all of which he with Humason tested to recession speeds roughly a tenth of the speed of light. These observations brought homogeneity to mind, but the evidence was not compelling.

The cosmological principle became an empirical construction by 1980, when we could add the evidence from the isotropy of the seas of X-rays and microwaves, and the test of scaling of the galaxy correlation functions with the depths of the galaxy catalogs.

We can imagine an alternative history in which de Sitter told Einstein about the astronomical evidence of a clumpy universe in January 1917, in time to persuade Einstein not to publish the argument for homogeneity. In this history the community settled on the natural alternative, Charlier's clustering hierarchy, which is Mandelbrot's fractal universe. The Shapley and Ames (1932) map fits this picture. The degree of clustering (the fractal dimension) could be chosen so Newtonian gravity is a good approximation on all scales, meaning the equilibrium velocity dispersion is the same at all levels of the hierarchy.[8] This is interestingly close to the observation that velocity dispersions of stars in galaxies and galaxies in the great clusters of galaxies are not very different. But the evidence Oort (1958) reviewed seemed puzzling, and two more puzzles were the smooth angular distribution of the radio sources and the close to isotropic X-ray sky. Surely the X-rays are from galaxies. But in a scale-invariant clustering hierarchy the galaxy distribution is distinctly clumpy on all scales, and in this picture the X-ray sky is similarly clumpy, contrary to the observations. The accumulating evidence was that the distant radio sources are galaxies. Their smooth distribution across the sky also is contrary to what is expected in a scale-invariant fractal universe. These observations forced attention to turn to homogeneity. The alternative history seems realistic, and the real history was just as messy. It shows how ideas can be corrected by the growing weight of the evidence.

---

8. In a scale-invariant fractal mass distribution the average value of the mass within distance $r$ of a particle of the mass is $M(<r) \propto r^D$, where the constant $D$ is the fractal dimension. The mean gravitational potential of the mass within distance $r$ of a mass particle is $U \sim GM(<r)/r \propto r^{D-1}$. If $D = 1$ then the gravitational potential can be small, Newtonian, on all length scales.

# CHAPTER FIVE

# The Hot Big Bang

The Second Word War drove intense effort in technological research, development, and manufacturing. In North America the end of the war released this great energy for peacetime science and technology, from relativistic particle accelerators to automobiles with fins. It does not seem surprising that some of this energy was directed to new thinking about the prewar ideas about gravity physics and cosmology presented in Tolman (1934) and Landau and Lifshitz (1951), which were close to sterile from an empirical point of view. Four main postwar actors, Robert Henry Dicke, George Gamow, Fred Hoyle, and Yakov Borisovich Zel'dovich, all excellent physical scientists, independently chose research aimed at enriching the physics of gravitation and cosmology, a Merton quadruple. In 1965 they and their research groups converged on recognition of the evidence that our universe expanded from hot dense early conditions that left two fossils: a large abundance of the element helium, and a sea of radiation at close to thermal equilibrium. (Radiation at thermal equilibrium has energy density at each wavelength determined by just one parameter, the temperature. The temperature unit, Kelvins, is the degrees of the centigrade scale but measured from the absolute zero temperature of thermodynamics.)

Gamow was an émigré from the Ukraine (as it was then termed), and spent the war years at George Washington University, Washington D.C. He knew nuclear physics; his book, *Structure of Atomic Nuclei and Nuclear Transformations*, was already in its second edition (Gamow 1937). He did not take part in the allied nuclear weapons program, perhaps because he was an émigré from the Soviet Union, maybe because he had little respect for authority. During the war Dicke in the US worked on the experimental side of electronics such as radar, and Hoyle in England worked on the applied side of radar. Zel'dovich contributed to the Soviet nuclear weapons program. He ended up at Arzamas-16, the USSR equivalent of the Los Alamos National Laboratory in the US. Rashid Sunyaev (2009) recalled that at Zel'dovich's "request librarians at Arzamas-16 had searched everywhere for all of Gamow's old papers." Since Gamow was an émigré this might have been dangerous, but Zel'dovich was named a Hero of Socialist Labour for his work on nuclear weapons. That gave him considerable influence over the bureaucracy, though his knowledge of the nuclear weapons program made it very difficult for him to leave the USSR.

Dicke and Zel'dovich established productive research groups. Dicke was a superb experimentalist. He liked theory, but he really cared about theory that has some relevance to interesting experiments or other empirical evidence. He formed the Gravity Research Group in about 1957. The contributions to the tests of gravity physics and the establishment of general relativity are discussed in Section 3.4. Most of the group's research has been on the experimental/empirical side, but Dicke had two theoretical graduate students: Carl Brans on the scalar-tensor theory of gravity (Brans and Dicke 1961), and me. Gravity naturally led Dicke to cosmology, and Dicke took me with him.

Zel'dovich's group was theoretical, though they paid close attention to developments on the empirical side. Sunyaev mentioned Zel'dovich's early interest in Gamow's ideas. Publications by Zel'dovich and his group on this and many other aspects of theoretical relativistic astrophysics and cosmology began in the early 1960s. The group remained productive until it was scattered by the breakup of the Soviet Union in the years around 1990.

## 5.1 Gamow's Hot Big Bang Cosmology

Gamow published two papers on cosmology in 1948 that are prime examples of his deeply imaginative and intuitive approach to physical science. He was guided in part by the pre-war thinking that the chemical elements might have been been formed by thermonuclear reactions among atomic nuclei, maybe during the hot early stages of expansion of the universe.

Thermal production of elements would require temperatures that von Weizsäcker (1938) estimated to be on the order of $10^{11}$ K, far hotter than the interiors of stars. He mentioned the observations of spiral nebulae, the galaxies, that seemed to be moving away from us as rapidly as a tenth of the speed of light. It suggested to him that (in my Google-aided translation)

> if one may interpret the red shift in the spectra of the spiral nebulae as a Doppler effect, then the extrapolation back to an explosive movement offers a concrete reason to ascribe to the world an essentially different physical state than today for a point in time approximately $3 \cdot 10^9$ years ago.

This is a first approximation to the hot big bang picture. (The relatively short expansion time, 3.9 billion years, results from Hubble's overestimate of the expansion rate.) Von Weizsäcker might have been thinking that a violent "explosive movement" would be expected to have been hot, maybe reaching temperatures sufficient to produce thermonuclear reactions. He might have known the relativistic big bang theory presented in Tolman (1934), and Tolman's demonstration that the homogeneous expansion of a universe filled with a sea of thermal radiation would cool the radiation while preserving its thermal spectrum. The thermal radiation is directly relevant, because the theory von Weizsäcker used to find predicted isotope abundance ratios to be compared to observed values assumes thermal (that is, statistical) equilibrium with a sea of thermal radiation.

The theoretical expression for these abundance ratios from statistical mechanics, the quantum theory of heat, is known as the Saha equation, after the Indian physicist Meghnad Saha. It applies to the process in which an atomic nucleus may gain or lose atomic weight by the capture or loss of free neutrons accompanied by the emission or capture of photons. The Saha equation describes the statistical equilibrium of capture and loss when the photons are a thermal sea of radiation. The important free parameter in Saha equilibrium is the radiation temperature. In von Weizsäcker's picture of the origin of the elements, the temperature during element formation is to be adjusted to get computed and measured abundance ratios to agree. Different chemical elements indicate different temperatures, but one can imagine ways to work around that.

Von Weizsäcker (1938) mentioned communications with George Gamow on the nuclear reactions that convert hydrogen to helium, catalyzed by the nuclei of carbon, nitrogen, and oxygen This is a path to the release of nuclear binding energy in stars that keeps them shining. It is possible then that Gamow knew von Weizsäcker's thinking about element formation. In the year following von Weizsäcker's paper, Gamow with Edward Teller published his first paper on cosmology, on an aspect of galaxy formation that is related to element formation in a hot universe, as follows.

Gamow and Teller (1939) considered the gravitational assembly of a bound concentration of mass such as a galaxy in an expanding universe. The matter temperature sets the minimum value of this mass, set by the condition that the attraction of gravity can overcome the resistance of the matter pressure. They proposed that the value of this minimum mass might account for a characteristic mass of a galaxy. Gamow and Teller referred to Jeans (1928), who derived this minimum mass under the assumption of a static universe in Newtonian physics. Jeans' calculation is easily adjusted to take account of the expansion of the universe in general relativity physics, but it was typical of Gamow to chose a shorter intuitive physically sensible route to essentially the same answer. The Gamow and Teller condition that a region can break away from the general expansion and form a gravitationally bound mass concentration is that

> the gravitational potential on the surface is larger than the proper kinetic energy of its particles.

This sets the Jeans length apart from an uninteresting numerical factor. After the war Gamow returned to thoughts in this direction, and the relation to element formation.

We can expect that von Weizsäcker, and Chandrasekhar and Henrich (1942), who explored von Weizsäcker's considerations in more detail, understood that their computations from statistical mechanics assume statistical equilibrium with a sea of thermal radiation. This is standard physics. There are some interesting details they do not seem to have noticed. They implicitly imagined a spatially uniform radiation temperature. In the spatially uniform big bang cosmology the sea of thermal radiation would not go away as the universe expanded. If the universe is the same everywhere—we have abundant checks of that now, and cosmologists usually assumed it then—the thermal radiation is close to the same everywhere; there is no other place for it to go.[1] The expansion of the universe would have lowered the radiation temperature while preserving the thermal spectrum, as Tolman had shown. The idea that the universe now contains a sea of thermal radiation, a cooled fossil left from the hot early universe, proves to be right; this fossil is detected at microwave wavelengths.[2] The idea was there to be recognized before the war, but no one seems to have noticed.

The large mass density in the very early universe would have caused rapid expansion, if you accept standard gravity physics.[3] Gamow (1946) pointed out that

> we see that *the conditions necessary for rapid nuclear reactions were existing only for a very short time*, so that it may be quite dangerous to speak about an equilibrium-state which must have been established during this period.... Thus if free neutrons were present in large quantities in the beginning of the expansion, the mean density and temperature of expanding matter must have dropped to comparatively low values *before* these neutrons had time to turn into protons. We can anticipate that neutrons forming this comparatively cold cloud were gradually coagulating into larger and larger neutral complexes which later turned into various atomic species by subsequent processes of β-emission [the emission of electrons that accompanies the conversion of neutrons into protons, which increases the atomic number].

1. Another picture that Arthur Milne (1932) and Oskar Klein (1956) considered is a violent explosion of a bounded mass in otherwise empty space. In this picture the thermal radiation would escape the bounds of the matter.

2. The microwave part of the electromagnetic radiation spectrum ranges from 1 mm wavelength to 1 m. The spectrum of the sea of fossil thermal radiation peaks up at wavelength about 2 mm, depending on how you define it, and is detected at wavelengths from 0.5 mm to about 1 m.

3. At mass density $\rho$ the expansion time in the hot big bang cosmology is on the order of $t \sim (G\rho)^{-1/2}$, where $G$ is Newton's gravitational constant. At high density $\rho$ the expansion time is short.

This idea of rapid neutron capture in the rapidly expanding early universe is a step toward Gamow's element buildup picture.

Gamow (1948a,b) recognized that the expansion of a hot universe would trace back in time to radiation temperatures so large that the radiation would have been hot enough, the photons energetic enough, to have decomposed all the atomic nuclei. That would mean the early universe contained a sea of free neutrons and protons. As the universe expanded and cooled, and the sea of photons grew less energetic, protons and neutrons could start to stick together and begin to form heavier isotopes of the chemical elements.

The first step in what would have been going on in element building is the two reactions

$$p + n \leftrightarrow d + \gamma. \tag{5.1}$$

The reaction going left to right is the capture of a neutron n by a proton p to form a deuteron d, the atomic nucleus of the stable heavier isotope of hydrogen. The binding energy released by the capture of the neutron is carried off by the $\gamma$-ray photon. In the reaction going from right to left a $\gamma$-ray photon with enough energy knocks the deuteron apart. At statistical, or thermal, equilibrium the rates going both ways are the same, and statistical mechanics gives the ratio of number densities of free neutrons to neutrons bound in deuterons, given the deuteron binding energy, the nucleon number density, and the temperature. This is an application of the Saha equation von Weizsäcker (1938) used, but to lighter elements.

Gamow's scenario builds on the Saha equation as follows. As the universe expanded and cooled through a temperature $T_c$ of about $10^9$ K, the rates of the reactions going the two ways in equation (5.1) would shift quite sharply,[4] from practically no deuterons at temperatures above this critical value $T_c$, to as many deuterons as would have had time to have formed while the universe continued to expand after the temperature fell below $T_c$. This means that, as

---

4. The sharp change at $T_c$ assumes the particle number densities are well below the number density of $\gamma$-ray photons. It is abundantly satisfied in the situation Gamow proposed.

the hot early universe expanded and cooled, element buildup would have commenced at a definite time set by when the temperature fell below $T_c$. This is an important point, one that Gamow seems to have been the first to recognize, in 1948.

Gamow took it that, once there was an appreciable accumulation of deuterons by the capture of neutrons by protons, the deuterons would readily fuse to form heavier isotopes, up to helium and maybe beyond. These element building processes tend to go faster than the production and dissociation of deuterons in equation (5.1). This is because the former involve rearrangements of protons and neutrons in atomic nuclei, while the latter requires the weaker electromagnetic interaction to create and annihilate photons. Gamow did not explicitly spell out all this, but it is the physical situation, Gamow knew physics, and what he did write agrees with the story. Anyway, through some combination of good luck and good management he hit on another key point, that the amount of buildup of isotopes of elements heavier than hydrogen would be determined by two things. First is the rate of capture of neutrons by protons with the production of photons. He knew this rate, from established nuclear physics. Second is the number density of neutrons and protons at temperature $T_c$. The greater this density the more complete the production of deuterium, and from there to the production of heavier isotopes by particle exchange reactions.

Gamow supposed there were comparable numbers of neutrons and protons, and he chose their number density at $T_c$ so the conversion to heavier elements would have left roughly half the matter as hydrogen. He stated that this agrees with what he took to be the fact that

hydrogen is known to form about 50 per cent of all matter.

As often, Gamow did not explain. He may have had in mind observations of the plasma ejected by stars to form planetary nebulae. The recombination lines of ionized hydrogen and singly ionized helium are observed in this plasma, and that allows estimates of the relative abundance of hydrogen and helium. (The abundances of heavier elements are much smaller.) Aller and Menzel (1945) reported hydrogen mass fractions in planetary nebulae of about 70%, with most of the rest helium, mass fraction $Y \sim 0.3$. (The notation is used often in this chapter: the hydrogen mass fraction is $X$, the helium mass fraction is $Y$, and the remaining mass fraction $Z$ is in heavier elements.) Aller and Menzel were summarizing pre-war measurements. Their helium abundances are close enough to what Gamow had in mind; maybe he knew them. Gamow also may have remembered Schwarzschild's (1946) estimate of the helium mass fraction in the sun, $Y - 0.41$, with almost all the rest hydrogen. This is mentioned in the book, *Theory of Atomic Nucleus and Nuclear Energy-Sources*, by Gamow and Critchfield (1949). In Schwarzschild's (1958) book, *Structure and Evolution of the Stars*, the helium mass fraction in the Milky Way is put at $Y = 0.32$, with the caution that this may be wrong by a factor of 2.

Gamow (1948a,b) only outlined the buildup of elements in his hot big bang picture. Fermi and Turkevich at the University of Chicago had the expertise in nuclear physics and the energy to work the computation. They confirmed Gamow's intuition, that the early stages of expansion of a hot relativistic universe, with Gamow's matter density at $T_c$, would be expected to have left about a third of the mass in the isotopes of helium, with almost all the rest in the stable isotopes of hydrogen. Fermi and Turkevich did not publish; Gamow (1949) presented their results and Gamow's associates Ralph Alpher and Robert Herman (1950) published the details.

Along with the considerable fossil abundance of helium in a relativistic hot big bang cosmology, there would be the remnant fossil sca of thermal radiation. Gamow (1948a,b) recognized that the radiation would still be present, cooler, after the formation of helium. He took the matter temperature to be the same as the radiation, and he used the matter temperature to compute the minimum mass that gravity can gather against the matter pressure. Gamow and Teller (1939) had considered this application of Jeans' (1902, 1928) calculation (discussed on pages 91 and 106), but they could only postulate the matter temperature. Gamow (1948a,b) had the temperature wanted for the conversion of an appreciable fraction of the hydrogen to heavier elements, so he had a definite estimate of the Jeans mass. He thought it would be relevant for the masses of galaxies. The general idea still looks interesting, but so far it has not been shown to be useful.

Although Gamow knew this fossil radiation would be present now, his impression was that it would be too cool to be detectable against all the radiation produced since the big bang in the course of cosmic evolution. Gamow's former graduate student, Alpher, and their close colleague, Herman, nevertheless took the bold step of estimating the expected present radiation temperature. Alpher and Herman (1948) reported that the temperature "is found to be about 5° K." This is remarkably close to the measurements obtained a decade and a half later. The complicated story of how they arrived at this temperature is reviewed in Peebles (2014).

## 5.2  The Steady-State Cosmology

In the year that Gamow introduced the hot big bang cosmology, Herman Bondi, Tommy Gold, and Fred Hoyle introduced the steady-state cosmology (Bondi and Gold 1948; Hoyle 1948). It assumes the universe is homogeneous, apart from local fluctuations, and expanding, consistent with the observed galaxy redshifts, at a rate of increase of separation that is proportional to the separation, which is Hubble's law. Matter is assumed to be continually spontaneously created, and the new matter is expected to collect by gravity in concentrations that become galaxies. These young galaxies would replace the older ones as they moved apart, keeping the universe in a steady state.

The idea of continual creation of matter was just made up, of course, but one can lodge the same complaint against the relativistic hot big bang cosmology. We have seen the sparse empirical support for the general theory of relativity in the 1940s, yet Gamow in 1948 proposed to follow Einstein in applying the theory to the immense scales of the observable universe.

The great practical merit of the steady-state theory was that it predicts the relations among galaxy counts, redshifts, and apparent magnitudes (the observed flux of light from a star or galaxy). These predictions offered fixed targets for the observational tests of cosmological models that were under discussion then. The relativistic big bang model cannot predict these relations because of the light travel time. We see distant galaxies as they were in the past, when the numbers of galaxies and their luminosities would be expected to have been different from the older galaxies seen around us. The numbers and luminosities of galaxies would have been the same in the past if the universe were in a steady state, of course.

Gamow's hot big bang version of the relativistic expanding universe introduced by Friedman (1922, 1924) and Lemaître (1927) does have testable predictions: the fossil thermal radiation and the considerable abundance of fossil helium.

## 5.3   Fossils from the Big Bang: Helium

Hoyle was a firm advocate of the philosophy of the steady-state cosmology and skeptical about the idea of Gamow's hot big bang. But he was among the first to publish recognition of one of the two major signatures of Gamow's theory, a much greater abundance of helium than would be expected from production in stars, and in an amount comparable to what would be expected in Gamow's cosmology.

Hoyle made important contributions to the theory of formation of the chemical elements in stars. This was in part motivated by his distrust of the evolving big bang cosmology in which elements would have been formed in the hot early stages of expansion of the universe. Hoyle's logical alternative was stars. He collaborated in the assembly of arguments for how element formation in stars would go, presented in the influential paper by Burbidge, Burbidge, Fowler, and Hoyle (1957) that is known as $B^2FH$.

A leading hint in the direction of the $B^2FH$ considerations was the observation that older stars have lower abundances of the chemical elements heavier than helium. That would be the case if the oldest stars had formed when the stars were not very far along in their production of the elements. The abundance of helium is difficult to measure in stars, because the large excitation potential of helium causes formation of spectral lines that can only be observed in the hot surfaces of massive stars, and interpreting these data is difficult. We have noted that helium can be measured in interstellar plasma, however.

Hoyle's (1958) thinking about helium is indicated by a discussion recorded at the 1957 Vatican Conference on Stellar Populations:

**Hoyle:** The difficulty about helium still remains, however.

**Martin Schwarzschild:** The evidence for the increase in heavy elements with the age of the galaxy supports [the idea of element formation in stars]. However, it does not necessarily mean that He production occurs mainly in stars. Gamow's mechanism may work up to mass 4.

**Hoyle:** That is why a knowledge of the He concentration in extreme population II is so important.

The old stars in Population II have low abundances of the heavier elements. We have seen that Gamow's hot big bang picture naturally puts about a third of the mass in helium before stars started forming, leaving almost all the rest hydrogen with only slight traces of the heavier elements. This picture would predict that the old Population II stars have mass fraction $Y \sim 0.3$ in helium, with almost all the rest hydrogen. The exchange between Hoyle and Schwarzschild (1958a) shows that, in 1957, they understood that determining the helium abundance in old stars with low abundances of the heavier elements would offer a fascinating cosmological test.

Hoyle (1949) had argued earlier that

> in spite of the steady conversion of hydrogen to helium and higher elements, taking place within the stars, which amounts to the transformation in $10^9$ years of about 0.1 percent of the hydrogen present in the nebulae, hydrogen still constitutes about 99 percent of all material.

This of course assumes our Milky Way galaxy formed out of close to pure hydrogen, which might be expected of the continual creation of matter in the steady-state cosmology. The prediction is different in the hot big bang picture that Gamow presented the year before, with its early production of helium.

In the years around 1960 the astronomy community was beginning to realize that the helium abundance in our Milky Way galaxy is large, maybe in line with Gamow's ideas. Schwarzschild (1958b) put the helium mass fraction in the Milky Way at about $Y = 0.3$, give or take a factor of 2. Geoffrey Burbidge (1958) reviewed the astronomical evidence that the helium abundance is at least $Y \sim 0.1$ in the Milky Way. He argued that the stars are not likely to have converted this much hydrogen to helium and heavier elements, and discussed possible explanations, but Burbidge did not mention that Gamow's theory could account for the helium. Burbidge attended the 1953 Michigan Symposium on Astrophysics, where Gamow (1953a) discussed his hot big bang picture. Burbidge may have forgotten, or may not have attended the lecture, or may not have taken the idea seriously.

In the second edition of his book, *Cosmology*, Bondi (1960, p. 38) added the remark that the helium abundance may be larger than can be understood by formation in stars, and "might be explained on a cosmological basis." This could be a hint to Gamow's hot big bang theory; Bondi did not explain.

Donald Osterbrock (2009) recalled attending Gamow's (1953a) lectures at the Michigan Symposium, and remembered being impressed by Gamow's interesting remarks on a variety of subjects. As it happened, Osterbrock and Rogerson (1961) had a measure of the solar helium abundance from a model for the structure of the sun fitted to its mass, luminosity, and radius. This model has a free parameter, the abundance of heavy elements. The heavy elements matter because the radiation produced by nuclear reactions in the central parts of the sun is slowed in its diffusion out of the sun by scattering by the free electrons from the heavy elements. Osterbrock and Rogerson had an improved measurement of the solar heavy element abundance. With this, their solar model yielded a measure of the mass fraction of helium in the sun (because at given pressure and temperature the mass density of ionized helium is larger than that of hydrogen). They found helium mass fraction $Y = 0.29$.

Osterbrock and Rogerson saw that, within measurement uncertainties, their solar helium abundance agrees with the helium abundance in present-day interstellar matter in the Orion Nebula, a cloud of interstellar plasma ionized by hot young stars. And a similar abundance was seen in planetary nebulae.[5] It is suggestive

---

5. As we have remarked, the presence of helium in astrophysical plasmas is signaled by its recombination lines. The Orion Nebula is a cloud of interstellar matter that is ionized by the hot young stars that formed out of the matter in the cloud. The helium abundance in the plasma gives an indication of the present abundance of this element in interstellar matter. The star at the center of a planetary nebula has converted most of the matter in its core from hydrogen to helium and then fused the helium to carbon, nitrogen, and oxygen.

that, in contrast to the lower abundance of heavy elements in older stars, all these objects have a similar helium abundance, $Y \sim 0.3$. Osterbrock and Rogerson remarked that

> It is of course quite conceivable that the helium abundance of inter-
> stellar matter has not changed appreciably in the past $5 \times 10^9$ years,
> if the stars in which helium was produced did not return much of it
> to space, and if the original helium abundance was high. The helium
> abundance $Y = 0.32$ existing since such an early epoch could be at
> least in part the original abundance of helium from the time the uni-
> verse formed, for the buildup of elements to helium can be understood
> without difficulty on the explosive formation picture.[21]

Their reference 21 is to Gamow (1949).

Osterbrock and Rogerson (1961) published the first suggestion I have found of evidence of the presence of a fossil left from the hot early state of the universe: the large abundance of helium. Shortly after that Geoffrey Burbidge (1962) made the same point, in a review largely of the Burbidge, Burbidge, Fowler, and Hoyle (1957) arguments for the formation of the chemical elements in stars. In this paper Burbidge also reviewed the earlier considerations in Burbidge (1958) of the evidence of a large helium abundance in the Milky Way, added the evidence from Osterbrock and Rogerson (1961), and remarked that

> Since the material of the sun condensed about $5 \times 10^9$ yr ago, we need
> to explain both the comparatively high value of the He/H ratio as com-
> pared with the estimates [of production in stars] made earlier, and the
> fact that there has been little change in the relative abundance in the
> intervening $5 \times 10^9$ yr.

Burbidge offered possible explanations: maybe low mass stars have the expected low helium abundances; maybe the Milky Way produced the helium in an early very luminous phase,

> but in a steady-state universe such an hypothesis is not tenable, since
> we do not see any galaxies in this form today;

---

It is evolving to a white dwarf star. Helium in a planetary nebula, which came from the outer parts of the star, is not thought to have formed in the star, but rather to have to have been there when the star formed.

maybe the helium was produced in black giant stars (hypothetical stars lying deep inside optically opaque clouds); or maybe

> The bulk of the transmutation of H to He took place in the first few
> minutes in the expansion of the Universe.

Burbidge's references for this last proposal are Gamow (1953b) and Alpher and Herman (1950).

Burbidge's (1958) earlier paper on the helium abundance did not mention this hot big bang idea. Osterbrock and Rogerson (1961) introduced the idea, and it certainly looks interesting now, but this line of thought seems to have ended a year later at Burbidge (1962). Things of his sort happen; I can offer personal evidence of it. I published in Peebles (1964) a study of the structure of the planet Jupiter. It has element abundances similar to the sun: largely ionized hydrogen and helium. To match all the constraints, I had to postulate that the abundance of helium by mass is $Y \simeq 0.18$. To support the assumption of this large helium abundance I presented a list of astronomical estimates, including the Osterbrock and Rogerson (1961) paper. Their "explosive formation picture" would have meant nothing to me then. What I cannot explain is that I forgot the Osterbrock and Rogerson argument for a large helium abundance the following year, when I had turned to the analysis of helium formation in a hot big bang, independent of Gamow. Research in science tends to resemble a random walk, loosely guided.

O'Dell, Peimbert, and Kinman (1964) presented more evidence of large helium abundances in old stars, from observations of the globular star cluster M 15. They reported that

> The results of a photographic spectrophotometric investigation of the
> nebulous object K648 in the globular cluster M15 are presented....
> The oxygen abundance, relative to hydrogen, is deficient by a factor of
> 61 relative to the Sun. The helium abundance, $N(\text{He})/N(\text{H}) = 0.18 \pm$
> $0.03$ [$Y = 0.42 \pm 0.04$], is compared with that found in field planetary
> nebulae, a very high-velocity planetary nebula, and the Orion Nebula.
> It is shown that strong arguments can be presented, supporting the
> hypothesis that the original helium content of the globular clusters was
> much higher than the very low values usually assumed.

The low abundance of oxygen indicates the stars in this cluster formed when element production by stars was just beginning. The nebulous object in the cluster is a planetary nebula with a large helium abundance. O'Dell et al. pointed to the similarly large helium abundances in the Orion Nebula and in other planetary nebulae. They did not mention the large helium abundance in the sun, but they referred to Osterbrock and Rogerson, who had developed that measure. Robert O'Dell, in a private communication, recalls that they were not aware of Gamow's prediction of a large helium abundance produced in the early universe. Osterbrock and Rogerson (1961) mentioned it, but they were not very emphatic, and Burbidge (1962) mentioned it, but in a brief remark in a long review of element formation in stars.

The evidence collected in *Finding the Big Bang* (Peebles, Page, and Partridge 2009) is that the O'Dell et al. paper led Hoyle to give serious thought to the evidence that the abundance of helium is larger than can be expected to have been produced in the known kinds of stars, and that the helium might have been produced in Gamow's hot big bang. Hoyle's paper with Roger Tayler, with the title *The Mystery of the Cosmic Helium Abundance*, pointed out this interesting situation. Their arresting title was likely to capture attention, and they published in a more visible journal, *Nature*. It hard to imagine that their paper, and the idea of fossil helium, would have gone unnoticed for long if the fossil radiation had not been noticed at close to the same time.

Hoyle and Tayler (1964) pointed to another possible interpretation, that explosions of supermassive stars also could produce a lot of helium. A large galaxy typically has at its center a compact object, very likely a black hole, with mass that may range from $10^6$ to $10^{10}$ solar masses. Matter falling into these massive compact objects produces explosions. The origin of these objects is debated; one might imagine they are remnants of Hoyle and Tayler's supermassive stars. But as Burbidge (1962) pointed out, this would not be in the spirit of the 1948 steady-state cosmology, because these little bangs are not observed in the galaxies around us. It would be understandable in the big bang theory, because nearby galaxies are old, presumably past the stage of supermassive stars. But in the steady-state theory there are supposed to be newly formed galaxies nearby, and if exploding supermassive stars were common we ought to have seen examples.

Meanwhile, in the Soviet Union, Zel'dovich knew Gamow's hot big bang theory, and thought it had to be wrong because he was under the impression that old stars contain little helium along with low abundances of the heavier chemical elements. He asked Yuri Smirnov, a member of his group, to check the computation of element production in Gamow's theory. Smirnov (1964) found that, if the matter density is comparable to astronomers' estimates, then the helium mass fraction coming out of the hot big bang is $Y \sim 0.3$, a value that has become familiar in this story. At mass density one percent of the astronomers' estimate, deuterium would form in the hot big bang but the baryon density would be so low that little of the deuterium would fuse to helium. Smirnov found that the result would be helium mass fraction $Y \sim 0.05$, which might have satisfied Zel'dovich, but at this low mass density the residual abundance of deuterium would be unacceptably large. (We must now state the mass density in baryonic matter, because the evidence is that a considerable mass fraction is dark matter that would not take part in nuclear reactions. But let us ignore this detail for the moment.)

Smirnov concluded that

> The theory of a "hot" state for prestellar matter fails, then, to yield a correct composition for the medium from which first-generation stars formed.

Smirnov found what are now known to be reasonable pre-stellar abundances of the isotopes of hydrogen and helium when he used the astronomers' estimate of the mass density. Smirnov's computation is essentially the same as the one I was doing, independently, at about the same time, as will be discussed. But he had the wrong information about the abundance of helium. I had the right information at one time, but had forgotten it, and in early 1965 felt open to the possibility of large or small pre-stellar helium abundance. (I consulted two local experts, Martin Schwarzschild and Bengt Strömgren, about the helium abundance. They were polite but reserved. I now suspect they felt I should make my own decision by looking into the literature.)

The apparent failure of Gamow's hot big bang theory led Zel'dovich (1962) to consider the alternative, a cold big bang. He saw the problem. Suppose the early universe was as cold as possible, temperature $T = 0$, with matter at its ground level of energy. That would have been an elegant early condition, except that electrons satisfy the exclusion principle, so pushing them together would have raised their minimum energy, what is known as their degeneracy energy. At really high density, in very early stages of expansion of the universe, the electrons would have been energetic enough to force themselves onto protons to turn them into neutrons. The problem with this is that, as the universe expands and the density decreases, the degeneracy energy of the electrons decreases. This allows neutrons to decay to protons, releasing electrons, because room has opened for them. The new protons readily combine with other neutrons to produce deuterons that fuse to heaver elements. This universe ends up with little hydrogen, which is unacceptable because hydrogen, with the proton as its atomic nucleus, is the most abundant element. Zel'dovich's solution was to postulate that, in the very early stages of expansion of a cold universe, there were equal number densities of protons, electrons, and neutrinos. As before, the high densities in the early universe would have forced the electrons to large degeneracy energies. But it would have forced the neutrinos to have even larger degeneracy energies, because electrons share the energy in two spin states, while neutrinos have only one. This prevents the electrons from combining with protons to form neutrons, because that process creates neutrinos, and their large degeneracy energy left no room for more of them. This neat arrangement would result in a universe of pure hydrogen before stars started forming the elements. But astronomers were starting to recognize the evidence that the helium abundance was large before the stars formed, as expected in Gamow's universe. Zel'dovich had an elegant theory but the wrong information.

## 5.4  Fossils from the Big Bang: Radiation

### 5.4.1  UNEXPECTED EXCITATION OF INTERSTELLAR CYANOGEN

The first hint to the presence of the other fossil, the sea of thermal radiation, came from the detection of the molecule cyanogen in interstellar space. Cyanogen, CN, is a chemically bound carbon and nitrogen atom. Recall that an isolated cyanogen molecule can only be in discrete levels of energy; this is quantum physics. It means a cyanogen molecule in its ground level can be excited to its first more energetic level by the absorption of a photon only if the photon has just the right energy. That means just the right wavelength. (Remember the photon energy is $h\nu$ at frequency $\nu$, wavelength $\lambda = c/\nu$, where $h$ is Planck's constant.) So interstellar cyanogen absorbs light from a star at the wavelength of the photons that have just the right energy to be absorbed by a cyanogen molecule in the ground level. The excited cyanogen promptly decays back to the ground level by emitting a photon in some other direction. The result is a sharp absorption line in the spectrum of light from a star. That absorption line, at 2.64 mm, reveals the presence of cyanogen along the line of sight.

The curious thing noticed in the 1940s was that the spectra have another line caused by absorption by cyanogen that is in its first excited level. Why is an appreciable fraction of cyanogen present in the first excited level? The cyanogen could be excited by a sea of radiation at wavelength 2.64 mm that would be absorbed by cyanogen in the ground level and raised to the excited level. If the radiation had a thermal spectrum then at equilibrium the ratio of numbers of cyanogen molecules in the excited and ground levels, by the absorption and emission of radiation at this wavelength, would be set by the radiation temperature. Andrew McKellar (1941) converted the observed ratio of depths of the absorption lines from the ground and first excited levels to the ratio of numbers of cyanogen molecules in the ground and excited levels, and from that to the effective radiation temperature. McKellar found about 2.3 K.

This temperature assumes the cyanogen is excited by thermal radiation. Collisions might have done it, though the needed collision rate seemed too rapid for interstellar space. And if the excitation is by radiation the spectrum need not be thermal; all that is wanted is the intensity of the radiation at wavelength 2.64 mm. That is, McKellar had found an effective or excitation temperature at 2.64 mm. This may be what Gerhard Herzberg (1950) had in mind when he wrote that McKellar's temperature "has of course only a very restricted meaning." But two years earlier Alpher and

Herman (1948) had estimated that the present temperature of the fossil radiation in Gamow's picture would be about 5 K. Within the uncertainties of the calculation of these two temperatures, from the cyanogen absorption lines and Gamow's hot big bang theory, they are consistent with the measurement of the radiation temperature made a half century later, $T = 2.725$ K. It is clear now that the cyanogen is excited by fossil radiation from the early universe.

In a review of the hot big bang theory presented in an appendix in the Gamow and Critchfield (1949) book, *Theory of Atomic Nucleus and Nuclear Energy-Sources*, Hoyle (1950) used the detection of interstellar cyanogen as evidence against Gamow's theory. He wrote that a hot big bang

> would lead to a temperature of the radiation at present maintained throughout the whole of space much greater than McKellar's determination for some regions within the Galaxy.

Hoyle did not explain further. If Gamow saw this review of his ideas, did he understand the comment about "McKellar's determination?" This piece of astronomical lore may have been quite enigmatic to the physicist Gamow.

Hoyle (1981) offered recollections of a discussion with Gamow of the issue of the cyanogen excitation temperature.

> I recall George driving me around in the white Cadillac, explaining his conviction that the Universe must have a microwave background, and I recall my telling George that it was impossible for the Universe to have a microwave background with a temperature as high as he was claiming, because observations of the CH and CN radicals by Andrew McKellar had set an upper limit of 3 K for any such background. Whether it was the too-great comfort of the Cadillac, or because George wanted a temperature greater than 3 K, whereas I wanted a temperature of 0 K, we missed the chance of spotting the discovery made nine years later by Arno Penzias and Bob Wilson.

Gamow was visiting the corporation General Dynamics, which would have provided the white Cadillac convertible car, or the funds for its purchase. In the years after the Second World War, industry and the military sought the presence of accomplished physicists, I expect in the hope that they may drop hints to what curiosity-driven research would come up with next.

When Gamow talked to Hoyle, in about 1956, what radiation temperature did Gamow have in mind? Gamow (1953b) mentioned $T = 6$ K, but by this time Gamow, who was not a well organized person, had replaced his physically correct 1948 considerations in favor of a postulate that cannot be justified, apart from the fact that it offers a similar present radiation temperature and the quite low mean mass density Gamow consistently assumed, not on very secure grounds.[6]

### 5.4.2 DICKE'S QUEST

At Princeton University in New Jersey, professor Robert Henry Dicke had his own reason to wonder whether the big bang might have been hot. It grew out of his search for a better empirical basis for gravity physics. His thinking expressed in the introduction to *The Theoretical Significance of Experimental Relativity* (Dicke 1964) includes the remark that research

> which could be easily rocketed into first place of importance by a significant discovery, is the general area of cosmological effects. These concern a wide variety of subjects: possible gravitational or scalar waves from space, the effects of a quasi-static scalar field generated by distant matter (if such a field exists), the continuous creation cosmology (which should be either laid to rest or else cured of its ills), Schwarzschild stars if they exist (a Schwarzschild star is one for which the star's matter is falling down the gullet of the Schwarzschild solution), massive generators of gravitational radiation, the global aspects of the structure of the universe, and the origin of matter. All these and many more are subjects concerning effects on a cosmological scale. These effects, if they could be observed, would give information about the most fundamental of the problems of physics.

---

6. Gamow's postulate is that there was an epoch during the course of expansion of the universe when there were equal contributions to the expansion rate equation (in the footnote on page 176) by the mass density in matter, the mass density in radiation, and the space curvature term. There is no rational justification for this assumption.

This is dated February 1964. Bob Diche's "gullet of the Schwarzschild solution," also known as a black hole, was a prediction; we now have good evidence that a black hole with mass several million times that of the sun is at the center of our galaxy. Bob was fascinated by the technology that would have to be developed to build a detector of gravitational waves. They have been detected, and the detections have shown us such remarkable things as the merging of massive black holes. Later in 1964 Bob charged me to consider the thought that space might be filled with thermal radiation left from the early hot stages of expansion of the universe, and he directed two other members of his group, Peter Roll and David Wilkinson, to construct a microwave radiometer, a device Bob had invented, to look for this radiation. Not long after that we had the first news of its detection, at the Bell Telephone Laboratories in New Jersey.

Dicke was thinking that the universe might be expanding from a bounce after a previous cycle of expansion and collapse. The universe in the last cycle would have contained stars, their starlight coming from the energy released by fusing hydrogen to heavier elements in the cores of stars. This starlight would have been squeezed as the universe contracted, maybe growing hot enough for thermal dissociation of the heavy elements produced by stars while the universe was last expanding and then collapsing. That would convert the heavy elements into a fresh supply of hydrogen for our cycle. In the language of thermodynamics, this conversion would have been a seriously irreversible process that produced the entropy of our hot universe. Never mind the cycle before that previous one, or the singularity theorems that were starting to appear and argue against the prospect of a bounce. Bob Dicke didn't find such considerations as interesting as the prospect of a measurement to explore a speculative but just possibly significant idea. If he had heard of Gamow's theory he had forgotten. It happens: we had to remind Bob that he had already used the microwave radiometer he invented during war research to set an upper bound $T \lesssim 20$ K on the possible temperature of a sea of thermal radiation (Dicke, Beringer, Kyhl, and Vane 1946).

This bound on the radiation temperature came from Dicke's use of his microwave radiometer for exploration of the prospects for the wartime development of radar at shorter wavelengths, around

1 cm. A particular concern was the absorption of microwave radiation by atmospheric water vapor. Thermodynamics allows inference of absorption by measurements of emission, and Dicke's radiometer could detect microwave radiation emitted by water vapor. The intensity of this radiation increases as the antenna is tilted away from the zenith, because the antenna receives radiation from a longer path through the atmosphere. The detection gave the wanted information about absorption. The extrapolation to zero path length gave Dicke et al. the bound on the temperature of an external uniform sea of radiation.

Dicke took his radiometer to Leesberg, 40 miles from Orlando, Florida, to measure radiation from the atmosphere at wavelengths 1 to 1.5 cm. Florida was chosen for its high humidity, because the issue at hand was the effect of water vapor on radar signals. Measurements at a less humid site, one with less emission from the atmosphere, would have allowed a tighter bound on an external sea of radiation. I have seen it argued that the radiation that is present, amounting to $T = 2.725$ K, might have been detected then, given the motivation and investment of resources.

When Dicke invited me to think about the theoretical implications of a hot big bang cosmology my first task was to learn what cosmology is. I turned to the books by Landau and Lifshitz (1951) and Tolman (1934). Neither is a source for the modest phenomenological considerations at play in cosmology then; I learned about them as I went along. I did not know about Gamow or his ideas. The notion of Dicke's hot big bang brought to my mind a pressure cooker that exploded, scattering about bits of food in different states of being cooked. That led me to think about the rapid expansion and cooling of an initial really hot sea of plasma and radiation. And that naturally led me to work through the same computation that Yuri Smirnov was doing in the Soviet Union, at essentially the same time. Also at about the same time, in Cambridge, England, Hoyle was growing interested in the evidence of large abundances of helium in old stars, and the possible relation to Gamow's hot big bang. At his suggestion, John Faulkner computed the first stage of this element building process, the evolution of the relative abundance of neutrons and protons before they started combining to deuterons and fusing to make helium. The three independent computations, by Smirnov, Faulkner, and me, had different but related motivations. It was a Merton triplet.

I had to consider that Roll and Wilkinson might detect nothing. I saw the challenge to a cold early universe: catastrophic production of elements heavier than hydrogen (which I mentioned on page 119). In 1964 I hit on the same solution that Zel'dovich found in 1962, the postulate of a suitably large number density of neutrinos. I published only a brief comment about my considerations of a cold big bang, in Dicke and Peebles (1965), because news had reached us of possible detection of a sea of radiation by microwave receivers used for communications experiments at the Bell Telephone Laboratories. It is no surprise that Zel'dovich and I reached the same picture for a cold big bang, because the physical situation was clear. The surprising thing is that we were thinking about hot and cold big bangs at about the same time, independently: a Merton doubleton.

I of course also considered the possibility that Roll and Wilkinson would detect a sea of radiation, and even set a lower bound, about 10 K, on its temperature, provided there are not the extra neutrinos that could save a cold big bang. To understand this lower bound, recall that element buildup would have commenced when the temperature of the sea of radiation fell through the critical temperature $T_c \sim 10^9$ K. That is when deuterons could start accumulating. Suppose the present matter density is given. Then the lower the present radiation temperature, the earlier the time in the course of expansion of the universe that the temperature would have passed through $T_c$. The earlier that happened, the larger the matter density at $T_c$. The larger the matter density, the more complete the nuclear reactions that combine neutrons and protons to produce deuterons that fuse to helium. If the present radiation temperature were too low, then too much helium would have been produced, hence the lower bound on the present temperature. So I was off by a factor of 3. Anyway, I was not yet looking into astronomical measurements of the cosmic helium abundance as closely as I ought to have been.

It was natural to check the expected accumulation of starlight and radio radiation from the galaxies, to see whether it might be expected to interfere with the detection of a sea of thermal radiation at any temperature Roll and Wilkinson could hope to reach. I concluded that it was not likely to be a problem because the fossil radiation would be concentrated in a relatively narrow range of frequencies that would stand out above the broader spread of wavelengths of the accumulated radiation from stars and galaxies. Doroshkevich and Novikov (1964) in Moscow considered the same issue and came to the same conclusion. We did not know about each other; it was yet another Merton multiple.

### 5.4.3 UNEXPECTED RADIATION IN BELL MICROWAVE RECEIVERS

At the Bell Telephone Laboratories in Holmdel, New Jersey, some 40 miles by road from Princeton, receivers used for experiments in communication at microwave wavelengths recorded detection of more radiation than expected. It amounted to a few degrees Kelvin in excess of what the engineers could account for by adding up all the known sources of radiation in the receivers and what was expected to be entering the receivers through their horn antennae from local sources of radiation. To my knowledge, DeGrasse, Hogg, Ohm, and Scoville (1959a,b) published the first comparison of the radiation detected in the Bell experiments to the sum of what was expected from known sources. To make the sum of sources agree with the detected radiation they assigned 2 K to antenna "side or back lobe pickup." This would be radiation from the ground that made its way into the horn-reflector antenna. But these antennae were designed to reject radiation coming from the back and sides far better than that; this was a real anomaly. It proved to be McKellar's, Gamow's, and Dicke's radiation, at temperature $T = 2.725$ K.

Doroshkevich and Novikov (1964) in Moscow were the first to recognize the importance of these Bell measurements for cosmology. They knew the DeGrasse et al. (1959a) paper, but did not recognize the unrealistic assignment of ground noise pickup. They assumed that this experiment put an upper bound on the Gamow radiation temperature; their example took this temperature to be 1 K. They suggested that

Additional measurements in this region [of frequency] (preferably on an artificial earth satellite) will assist in final solution of the problem of the correctness of the Gamow theory.

Indeed, measurements from a satellite and a rocket determined that the radiation detected by the Bell experiments has the thermal spectrum Gamow and Dicke expected of this fossil from a hot big bang.

In 1963 Arno Penzias and Robert Wilson, both new to the Bell Radio Research Laboratory at Crawford Hill, New Jersey, began a thorough search for the source of the excess microwave radiation detected in the Bell receivers. They did the right thing by persisting in the search, and the Bell management is to be credited for allowing them to do so. Penzias and Wilson also did the right thing by complaining about the problem until someone heard and directed them to the people at Princeton, who were looking for what the Bell receivers had already detected, fossil radiation from the hot big bang.

The recognition that the excess noise in the Bell receivers might be what Roll and Wilkinson were looking for grew out of a lecture I presented on my research in cosmology at the Johns Hopkins Applied Physics Laboratory. That was in Maryland, on 19 February 1965. It only occurred to me many years later that Gamow's associates Ralph Alpher and Robert Herman had worked there in 1948, the year of Gamow's memorable papers on the hot big bang. Maybe someone there who knew about the old days heard that I was doing something along those lines; I never thought to ask. I did ask David Wilkinson whether he would be comfortable with my mentioning his experiment with Peter Roll, and my thinking about what their results might mean. I think I have an accurate memory of his reply: "no one could catch up with us now." We did not imagine that the experiment had already been done and an anomaly detected.

A friend from graduate school days, Ken Turner, attended my Johns Hopkins lecture. He was at the Carnegie Institute of Washington, a productive center for research in the natural sciences. Ken told a colleague, Bernie Burke. Burke knew the problem with excess radiation in the Bell receivers, and he advised Penzias to call

Dicke. Dicke, Roll, and Wilkinson returned from a visit to Penzias and Wilson in March 1965 with the conclusion that they had a good case for detection of a sea of radiation that had to be pretty smoothly distributed across the sky, because the excess noise in the Bell receivers was about the same wherever in the sky the antennae pointed. That is what would be expected of a close to uniform sea of thermal radiation left from the early stages of expansion of a close to uniform universe.

The Bell paper on the excess radiation, and the Princeton paper on the interpretation, were submitted in early May and published on 1 July 1965 (Penzias and Wilson 1965; Dicke, Peebles, Roll, and Wilkinson 1965). The Princeton paper referred to two papers from Gamow's group: Alpher, Bethe, and Gamow (1948) and Alpher, Follin, and Herman (1953). I had not yet recognized the central importance of Gamow's (1948a,b) papers, and it took me even longer to see that our first reference, the Alpher, Bethe and Gamow paper, is a failed first attempt. The details are in Peebles (2014).

The announcements of detection and interpretation were greeted with the expected mix of interest and skepticism. Indeed, it was difficult to defend the argument that a sea of thermal radiation had been detected from the measurement of its intensity at just the one wavelength of the Bell measurement, 7.4 cm. But we could point out in our interpretation paper that if the radiation were a thermal fossil from the early universe then the hot big bang theory would predict that the helium abundance is large, about 30% by mass, which is about what was observed. A theory that fits two puzzling and otherwise quite different observations certainly is worth close attention.

The theory became really interesting in 1990 when precision measurements showed that the radiation is very close to thermal. Recall that the intensity of thermal radiation at each wavelength is determined by a single quantity, the temperature. The universe now is close to transparent at microwave wavelengths, so there is no process that would relax a sea of radiation to the observed thermal state. In the hot big bang picture the sea of radiation would have been driven to thermal equilibrium by the rapid rate of relaxation in the hot dense conditions in the early universe.

The expansion of the universe would cool the radiation but leave it close to thermal under three conditions. First, spacetime curvature fluctuations must be small, perhaps the minimum required for the gravitational assembly of galaxies and their spatial clustering. Larger departures from homogeneity could be significant, from what we knew then, but if so we would have been presented with a mix of thermal radiation at a range of temperatures, which is not thermal. Spacetime curvature around a black hole does greatly depart from the mean, and black hole absorption of radiation in principle makes the fossil sea depart from thermal, but the effects are tiny. The second condition is that the interactions of the sea of radiation with matter that is hotter or colder than the radiation, and with matter that is moving at high speeds in winds and explosions, have not seriously disturbed the fossil radiation. Third, it must be assumed that radiation from dust heated by starlight, and from galaxies that are radiating at microwave as well as radio wavelengths, do not get in the way. My estimates suggested these three effects would not be a problem, but estimates can be wrong.

An accurate measurement of the spectrum to check on the idea of a near thermal state, and indications of possible disturbances that caused departures from thermal, is difficult because it requires putting detectors above the atmosphere to avoid this local source of radiation. Reports of spectrum measurements in the two decades from 1970 to 1990 suggested there may be anomalies, departures from thermal, usually near the short wavelength end of the spectrum. The anomalies differed in different measurements, which made one think of systematic errors, but I had to consider that there could be a true anomaly. If so, the interesting challenge for inventive theorists would be to devise a picture for the evolution of cosmic structure that would account for this anomaly, and then find predictions that might test the picture. The interesting challenge abruptly changed in 1990 when two groups reported convincing demonstrations that the sea of microwave radiation has a very close to thermal spectrum. If the radiation has not been seriously disturbed on its way to us from the early universe, then the nature of its distribution across the sky might teach us about conditions in the early universe. That certainly attracted the attention of inventive theorists, and much became of it, as will be discussed in Section 6.10.

The origin of one of the programs that completed the spectrum measurement traces back to a 1974 proposal to NASA for a Cosmological Background Radiation Satellite. The name became COBE, for Cosmic Background Explorer. John Mather was designated the COBE Study Scientist and Principal Investigator for the spectrum measurement. The other program was led by Herb Gush, at the University of British Columbia. The two groups used the same detector technology (a Fourier transform interferometer). In the year that Mather and colleagues submitted their proposal to NASA, Gush (1974) reported an attempted spectrum measurement by a detector carried above the atmosphere by a rocket. It failed because Gush found that the weight limit allowed by the rocket did not permit adequate shielding from radiation from the earth. Sixteen years later, clear demonstrations that the spectrum is very close to thermal were announced by the NASA satellite group (Mather, Cheng, Eplee et al. 1990) and Gush's rocket group (Gush, Halpern, and Wishnow 1990). The first was submitted to *Astrophysical Journal Letters* on 16 January, the second to *Physical Review Letters* on 10 May of the same year. Considering the many years it took to get there, we may say the projects were completed at essentially the same time. Either experiment would have served for a convincing demonstration that the spectrum is close to thermal.[7] The independent demonstrations added weight.

The two 1990 measurements and later even more precise ones have not yet detected departures from a thermal spectrum. They must exist, because the universe is not at thermal equilibrium, and their eventual discovery will offer yet another interesting challenge: find an explanation for the departures from thermal that can be trusted because it passes tests of predictions.

The argument that the radiation had to have been thermalized in the early universe would fail if the universe as it is now were opaque, capable of absorbing and emitting radiation rapidly enough to force relaxation to thermal equilibrium. The idea was discussed. Hoyle and Wickramasinghe (1988) presented a detailed scenario for how

7. To avoid confusion let us note that, although there is no reason to doubt Planck's theory of the blackbody spectrum, the theory has not yet been well checked against measurements. The fossil radiation spectrum is demonstrated to be thermal because it is compared to sources at well-determined temperatures.

the right kind of intergalactic dust could make the universe opaque at microwave wavelengths, capable of relaxing radiation to thermal equilibrium, and yet transparent at optical wavelengths where galaxies at great distances are observed without serious absorption. The problem is that radiation from distant galaxies also is observed at microwave wavelengths, where the dust is supposed to be opaque and capable of thermalizing the sea of radiation. One might imagine that rifts in the dust clouds allow glimpses of some more distant galaxies, but that would make the radiation a mix of spectra, some less well thermalized at the rifts, which is not observed. It seemed clear, and it is even clearer now, that intergalactic space is close to transparent at microwave as well as optical wavelengths. The case that the sea of microwave radiation is a fossil from the early universe is about as persuasive as it can get.

It is remarkable that we have compelling evidence of the presence of two fossils from a time when our universe was very different from now. Arno Penzias and Bob Wilson were awarded the 1978 Nobel Prize in Physics for their role in the discovery of the cosmic microwave background radiation. John Mather and George Smoot shared the 2006 Nobel Prize in Physics for the COBE satellite demonstration of the thermal spectrum of this cosmic radiation, which showed that the radiation has not been seriously disturbed by explosions or the like, and the detection of slight departures from an exactly smooth sea of radiation, which is a signature of the gravitational disturbance caused by the growth of cosmic structure. Nobel Prizes are awarded for distinctly important contributions, but it should be understood that decisions for awards have to be capricious because there are more truly great contributions than there are serious prizes. Herb Gush certainly qualifies. He received the respect of colleagues who understood his great achievement, but not the Nobel Prize. The identification of the fossil helium passed through too many hands for specific recognition, another capricious effect.

The demonstration of the thermal spectrum of the fossil radiation encouraged another thought. General relativity shows how to compute the disturbance to the spatial distribution of the sea of radiation by the gravity that pulled together the concentrations of mass observed in galaxies and their clumpy spatial distribution.

This assumes gravity pulled together these mass concentrations, not explosions, for example. It also assumes explosions have not seriously disturbed the radiation, and it of course assumes we can trust general relativity. But if all this were so then observations of the variation of the fossil radiation across the sky would open a window into the state of the early universe. The pursuit of this idea is the subject of the next chapter.

# CHAPTER SIX

# The ΛCDM Theory

Two great experimental programs to be described in Section 6.10 of this chapter produced the evidence that made the case for what is known as the ΛCDM theory of the large-scale nature of the universe. The CDM part stands for cold dark matter, a hypothetical substance that interacts little if at all with radiation and the matter we are made of. The symbol Λ represents a hypothetical constant that Einstein introduced, and came to despise. But cosmologists have learned to live with it. The story of how we arrived at this theory, and the case for these two hypothetical components, includes considerations of several issues, beginning with ideas about the nature of conditions in the early stages of expansion of the universe.

## 6.1 Initial Conditions

The present spatial distributions of galaxies and the fossil radiation carry information about how cosmic structure grew in the remote past. To see this, recall first a different situation. The flow of water in a pipeline is unstable to the development of turbulence. The size of the pipeline and the density, viscosity, and speed of flow of the water determine a characteristic doubling time during which any small departure from exactly uniform flow doubles, and then doubles again and again. This is an exponential growth of departures from smooth flow, and the inevitable onset of turbulence. The properties of the fully developed turbulence do not depend on the initial departures from smooth flow. In effect, turbulence "forgets" its initial conditions; they do not matter for the study of fully developed turbulence. The physics of fluids matters, of course, and it is a considerable challenge to find a satisfactory application of this physics to reliable analyses of the properties of turbulence.

The relativistic expanding universe is unstable to the growth of small departures from an exactly uniform mass distribution. The big difference from the flow of water in a pipeline is that the departure of the mass distribution from uniformity grows as a power of the time, not exponentially. This means that the present distributions of mass and fossil radiation carry a "memory" of initial conditions, which is a very different situation from turbulence. In particular, the interpretation of the observed cosmic structure—galaxies and their clustered distribution—requires consideration of initial conditions. In developing the now well-tested ΛCDM theory of the evolution of the universe and its cosmic structure we had the freedom to adjust initial conditions to fit the observations. This alone amounts to a just so story, of course. The evidence that it is more than that is the considerable array of successful predictions to be discussed in Section 6.10. And since the theory has been well tested there is the positive side to the situation, that we are learning about the nature of our initial conditions, the primeval state of the expanding universe.

How shall we characterize the nature of the primeval departures from a strictly homogeneous mass distribution in the very early universe? In the general theory of relativity the departures from exact homogeneity perturb spacetime curvature. This means the departures of the curvature from the mean are a measure of the mass distribution. An early thought from simplicity was that these spacetime curvature fluctuations are the same on all length scales. If so the primeval departures from uniformity are specified by the choice of just one parameter, the amplitude of the departures from uniformity. The condition is known as scale-invariance.[1]

1. In the jargon, the primeval departure from exact homogeneity is assumed to be a stationary Gaussian random process that has a power law power spectrum with power law index fixed so the spacetime curvature fluctuations are close to the same on all length scales. The spacetime curvature fluctuations we observe around us are small. This is a good thing, because large curvature fluctuations tend to become black holes. A

The argument from simplicity for scale-invariant initial conditions was presented by Harrison (1970), Peebles and Yu (1970), and Zel'dovich (1972). It happens to be close to what was measured a quarter century later. It is an example of the simplicity the world sometimes offers us.

## 6.2   The Curvature of Space Sections

In a homogeneous expanding universe we can define a three-dimensional space through four-dimensional spacetime by all the spatial positions at a chosen value of the time elapsed since expansion began at some very early and large density. The value of this initial density does not matter, provided it is very large, because the expansion would have spent very little time at very large density. And we need not pause to worry about what happened earlier than that. There are ideas but we have little evidence.

The three-dimensional space sections defined this way may satisfy Euclid's trigonometry: parallel straight lines in a space section never meet, the exterior angles of a triangle sum to 360°, and all that. If space sections are flat, spacetime still is curved, taking into account the time direction, if there is mass. This spacetime is said to be cosmologically flat, even though it is curved in spacetime.

Another possibility is that the three-dimensional space sections are closed, in analogy to the closed two-dimensional space on the surface of a sphere. Or the space sections may be open, in analogy to the surface of a saddle. Open and cosmologically flat universes in principle can extend to spatial infinity, but that need not be so in our universe. We only know that, at the greatest distances that the theory says can be observed, there are no signs of encountering an

---

tilt of the power law index that makes spacetime curvature fluctuations large on some very large scale would be an odd situation: why would the curvature of spacetime on the scales we can observe be small if the large-scale curvature fluctuations in general are large? A tilt to large curvature fluctuations on very small scales would imply a sea of small black holes. Maybe they would vanish by Hawking's black hole evaporation, if the tilt is only slight, so they are not too massive. Of course, the initial conditions need not have proved to be close to a power law, but we have been fortunate: only a small tilt from scale-invariance is required to fit the demanding tests at the time of writing.

edge to our homogeneous universe of galaxies and radiation. Maybe there is chaos much further away; we can't tell.

In general relativity, solutions to the evolution of a homogeneous universe may be classified by analogy to the escape speed of a rocket fired from the earth. A rocket fired at escape speed moves arbitrarily far away while always being slowed by the attraction of gravity. If the rocket is fired at less than escape speed it falls back to the earth, and if greater it escapes the gravity of the earth and moves away freely. (I am not taking into account all the other masses around us.) For application of this example to the evolution of the universe, suppose first that Einstein's cosmological constant is not present. Then if space sections are closed, as in a sphere, the universe is expanding at less than the equivalent of escape speed. The expansion eventually stops and the universe collapses to an analog of a black hole. Our theory is not advanced enough to say what would happen after that big crunch. An open universe expands at greater than escape speed, meaning it eventually escapes the diminishing attraction of gravity and the rate of expansion settles to a constant value. In the general theory of relativity as now formulated, and without the cosmological constant, this open universe continues expanding into the indefinitely remote future, a big freeze. A cosmologically flat universe also expands into the indefinitely remote future, but the rate of expansion keeps being diminished by the ever decreasing attraction of gravity.

A fascination with how the world will end, a big crunch or a big freeze, is understandable, I suppose, even though there is no need to make preparations. But I would not trust the theory we have now for an extrapolation into the remote future, because I am seriously impressed by the immense extrapolation of our standard physics in its application to the behavior of the observable universe up to the present epoch. The tests look good, but extrapolation to the indefinitely remote future is pressing our luck. Anyway, interest in how the world will end served the useful purpose of inspiring research that helped us reach the conclusion that the universe is cosmologically flat or very close to it; that there are the two kinds of matter, baryonic and dark, discussed in Sections 6.5 and 6.6; that there is the fossil sea of thermal radiation and neutrinos left from the hot early universe that was discussed in Chapter 5; and that there is the puzzling cosmological constant to be considered next.

## 6.3 The Cosmological Constant

Chapter 4 reviews Einstein's (1917) presumption that the universe is unchanging: static apart from minor local fluctuations. This seems natural enough, likely the first thing one might imagine. Einstein found that his general theory of relativity does not allow a static homogeneous universe of matter with positive or zero pressure. To remedy this he added a term to his gravitational field equation that has the effect of pushing matter apart: the greater the separation, the stronger the repulsion. The strength of the repulsion at a given separation is determined by the value of a new physical constant that came to be termed Einstein's cosmological constant, $\Lambda$. If this constant has just the right value then its repulsion balances the attraction of gravity, allowing a static universe. Einstein (1917) wrote that he introduced this "supplementary term" with the constant $\Lambda$ to his field equation "only for the purpose of making possible a quasi-static distribution of matter, as required by the fact of the small velocities of the stars."

A quasi-static cluster of stars in asymptotically flat and empty spacetime could have small velocities, consistent with what is observed. But we saw in Chapter 4 that Einstein was looking for an arrangement with matter everywhere, consistent with his thinking about Mach's principle.

Einstein did not notice that his balance of the attraction of gravity and the repulsion of the cosmological constant is unstable. In his static universe, a region in which the mass density happens to be slightly greater than the mean has a local gravitational attraction that is slightly greater than the mean, that is, greater than the repulsion of the $\Lambda$ term. The local density grows larger, exponentially. Even a tiny departure from the mean grows large after a few dozen doubling times. As we have seen in Section 4.1, James Jeans (1902) found this exponential instability in Newtonian physics. Essentially the same analysis applies to Einstein's (1917) static universe, but this was only recognized a decade later.

Einstein came to regret introducing the Λ term. In May 1923 he wrote to Hermann Weyl that[2]

> According to De Sitter, two material points sufficiently distant from each other accelerate as they move apart.... If there isn't any quasi-static universe after all, then get rid of the cosmological term.

Georges Lemaître, who was a leading actor in the discovery of the expanding universe (as pointed out on page 92), saw merit in the cosmological constant. He entered these considerations in his contribution to the essays in the book, *Albert Einstein, Philosopher-Scientist* (Schilpp 1949), and in advance of that he wrote about them in a letter to Einstein. The Archives Georges Lemaître[3] has copies of Lemaître's letter, dated July 30 1947, and Einstein's reply, dated September 26 1947. Here is the second paragraph in Einstein's letter:

> Since I have introduced this term I had always a bad conscience. But at that time I could see no other possibility to deal with the fact of the existence of a finite mean density of matter. I found it very ugly indeed that the field law of gravitation should be composed of two logically independent terms which are connected by addition. About the justi-fication of such feelings concerning logical simplicity it is difficult to argue. I cannot help to feel it strongly and I am unable to believe that such an ugly thing should be realized in nature.

We see how in physics intuition can be influential but not always reliable. The evidence is that must learn to live with this ugly thing.

Given the present rate of expansion of the universe, the exis-tence of a positive cosmological constant Λ increases the age of the universe, because Λ requires slower expansion in the past. That is, the presence of a positive Λ increases the rate of expan-sion, so to match the present rate the expansion had to have been slower in the past and the expansion time longer than if Λ were not present. Lemaître pointed this out in his 1947 letter to Einstein. It seemed important at the time, because the present expansion

2. Collected Papers of Albert Einstein Vol. 14, Doc. 40.

3. I am grateful to Liliane Moens for providing me with this and other information from the Archives Georges Lemaître, Université catholique de Louvain, Louvain-la-Neuve, Belgium.

rate, Hubble's constant, had been seriously overestimated, which made the universe without a positive $\Lambda$ seem dangerously young compared to ages from astronomy and geology. Einstein replied to Lemaître that

> in the shortness of $T_0$ [the age of the expanding universe] there is a reason to try bold extrapolations and hypotheses to avoid contradiction of facts.

But the following paragraph of the letter, which is copied above, shows Einstein was being polite.

Others agreed with Einstein's opinion of $\Lambda$. Wolfgang Pauli (1958, p. 220), in the Supplementary Notes to the English translation of his book, *Theory of Relativity*, wrote that Einstein

> completely rejected the cosmological term as superfluous and no longer justified. I fully accept this new standpoint of Einstein's.

In *The Classical Theory of Fields*, Landau and Lifshitz (1951, p. 338) state that

> Nowhere in our equations do we consider the so-called cosmological constant, since at the present time it has finally become clear that there is no basis whatsoever for such a change in the equations of attraction.

One might read some flexibility in this way of putting it; perhaps they were open to the chance that a basis will be discovered.

Lemaître (1934) offered an idea for a basis for the cosmological constant. He recognized that $\Lambda$ is in effect an energy density, with pressure equal to the negative of the energy density. This is not the familiar kind of fluid, which is violently unstable when the pressure is negative. The exception is a fluid that is the same everywhere and the pressure is exactly the negative of the energy density. $\Lambda$ acts like this odd kind of fluid. In the 1947 letter to Einstein, Lemaître pointed out that this interpretation of $\Lambda$ offers a way to define the zero of energy. In quantum physics energies are computed only up to an arbitrary additive constant. That is all that matters in, say, the computation of the binding energy of an electron to an atom, which is the change of energy when the electron is pulled out. In such exchanges of energy the additive constant cancels. But the total energy, which is mass, matters for gravity. Lemaître's point was that Λ defines the energy density of empty space, to which all other forms of energy are to be added. Maybe this bears some relation to the puzzle of the zero-point energy of quantum physics, as follows.

Consider a hydrogen molecule: two chemically bound hydrogen atoms. In classical physics the distance between the two atoms can oscillate, at a frequency that is very nearly independent of the amplitude of the oscillation. In quantum physics the energy of oscillation of the separation of the atoms has discrete, nearly equally spaced, values. The ground energy level is one half of the difference between levels. This zero-point energy has a perfectly definite physical meaning. The binding energy of the hydrogen molecule in its ground level is the work required to pull apart the two atoms and leave them well separated at rest. The quantum physics calculation of this binding energy must take account of the presence of the zero-point energy associated with the distance between the two atoms in the molecule, which of course is not present when the atoms are pulled apart. The computed binding energy agrees with the measured value only if the zero-point energy of the molecule is taken into account. That is, this zero-point energy is physical, real.

Jordan and Pauli (1928) wrote that this zero-point energy must be real for matter, because experiments support it, but they expressed doubts about applying it to the electromagnetic field. They wrote (in my Google-aided translation) that

> Various considerations seem to speak in favor of the fact that, in contrast to the natural vibrations in the crystal lattice (where both theoretical and empirical grounds speak for the existence of a zero-point energy), in the free vibrations of the radiation field the "zero-point energy" $h\nu/2$ per degree of freedom is not physically real. Since one is dealing with strictly harmonic oscillators, and since that "zero-point radiation" can neither be absorbed nor scattered or reflected, it seems to evade any possibility of detection, including its energy or mass. It is therefore probably the simpler and more satisfactory view that zero-point radiation does not exist at all in the electromagnetic field.

In his Handbuch der Physik article on quantum mechanics, Pauli (1933, page 250) repeated the Jordan and Pauli argument, and added that (in the English translation in Rugh and Zinkernagel 2002)

> it is more consistent from scratch to exclude a zero-point energy for each degree of freedom [of the electromagnetic field] as this energy, evidently from experience, does not interact with the gravitational field.

This argument is wrong. Just as for the crystal Jordan and Pauli mentioned, the energy or mass associated with the zero-point energy $hv/2$ of a hydrogen molecule is real. This is an experimental fact. The standard and successful application of quantum physics to the electromagnetic field is the same as for a crystal or a hydrogen molecule. Either we modify quantum physics, which would be a serous challenge at this level of evidence for the theory, or we accept the reality of the zero-point energy of the electromagnetic field.

Enz and Thellung (1960) recall that

> The zero-point energy of the radiation has a special aspect compared to that of the lattice vibrations: the problem of its gravitational effect. Pauli recognized this question early on and, as he told us with obvious amusement, made a calculation of the gravitational effect at a young age. The result was that the radius of the world (if the short wavelengths were cut off at the classic electron radius) "would not even reach the moon."

It certainly is understandable that Pauli felt the zero-point energy of the electromagnetic field ought not to be real. But why would the same physics make zero-point energies real only in conveniently chosen applications? If always real, and if local physics is independent of the motion of the observer, meaning in particular that the electromagnetic zero-point energy density is independent of the motion of the observer, then its gravity acts like Einstein's cosmological constant. It would be Lemaître's fluid, but the expected value of this energy density would be absurdly large.

We discussed this problem in the 1960s in Dicke's Gravity Research Group. It was and remains profoundly puzzling, a "dirty little secret" of cosmology. Others recognized it. Zel'dovich (1968) discussed the problem. Weinberg (1989) surveyed attempts to adjust the physics to rationalize the situation. The failures left Weinberg with two thoughts. One is the hope that some symmetry of a deeper fundamental physics to be discovered requires that the ground level of energy density vanishes. It would be the asymptotically approached vanishing energy density of our cosmologically flat universe as it continued to expand at an ever slowing rate into the indefinitely remote future. That would be the sum of field zero-point energies, latent heats of cosmic phase transitions in the early universe, and maybe an intrinsic value of the cosmological constant.

The other thought Weinberg considered is an anthropic argument. If the cosmological constant Λ were negative, and its absolute value large, then the initially expanding universe would have collapsed back to a big crunch before we could have evolved. We would not be here to worry about it. If Λ were positive and large, it would have caused the universe to expand too rapidly to allow gravity to form galaxies. Our existence requires the gravity of a galaxy for confinement of the debris formed and shed by stars, and cycled through a few generations of stars, to produce the chemical elements we are made of. We can think of this situation as a consistency condition: if the absolute value of Λ were much larger than observed then we would not be here to measure it. The original thinking in this direction is Dicke's (1961) point, that it is not surprising that we find ourselves in a galaxy that is several billion years old. This much time is needed for the formation of the chemical elements in a few generations of stars, their collection in the solar system, and on the earth the evolution of the species up to us. The condition for falsification of this weak form of the anthropic argument, the requirement of consistency, is straightforward, at least in principle. Either we see consistency or we do not.

Weinberg discussed a strong anthropic argument, which invites us to imagine a statistical ensemble of universes. The cosmological inflation picture to be discussed in Section 6.4 encouraged the thought that inflation might be eternal, and might have nucleated many universes to produce a multiverse. Maybe different universes have different values of the cosmological constant. Then we would flourish in one of the universes in this multiverse that has a value of Λ that allows us to exist. This idea is not so easy to assess, to falsify. For example, one might ask why there are so many galaxies in our universe; wouldn't one do? But maybe the formation of something like us is so exceedingly unlikely that a great many galaxies were needed to improve the chance of our existence, which we presume is real. Some accept the strong form of the anthropic argument as reasonable and unavoidable; others dismiss it as a just so story.

In the 1970s and earlier there was not a community position in astronomy and cosmology on whether the cosmological constant term is worth considering. We have seen that Lemaître (1934) argued for it. But, at a time when the reach of the cosmological tests was quite limited, a practical consideration was that if the cosmological constant is ignored it reduces the number of parameters to consider with the modest data on hand (Peebles 1971). McVittie's (1956) position was that if $\Lambda$ vanished it would be "a numerical fluke whose occurrence would be very remarkable." His point was that, in general relativity and given Hubble's constant, the vanishing of $\Lambda$ places an identity relating the curvature of space sections and the mean mass density. McVittie felt that the measurements of these two qualitatively very different phenomena are not likely to satisfy this identity. Others implicitly agreed with McVittie, by considering the cosmological constant to be just another parameter to be measured, along with Hubble's constant, space curvature, and the mean mass density. These are four quantities to be adjusted to fit the observations, with the one constraint imposed by general relativity. For example, Petrosian, Salpeter, and Szekeres (1967) considered how the value of $\Lambda$ would affect counts of quasars as a function of redshift. Gunn and Tinsley (1975) considered the constraint on $\Lambda$ from combined estimates of the age of the galaxies, the mean mass density, the rate of expansion of the universe, and the rate of change of the expansion rate. Gunn and Tinsley commented that

> We seem to be in the situation once again where the data may call for $\Lambda$ to be dusted off and inserted in the field equations.... Most relativists find it repulsive in principle rather than by observation. If it is regarded as a fundamental constant of the classical theory, it does indeed reduce the lustre of an otherwise beautiful framework.... The first reaction to all this [evidence of detection of the effect of $\Lambda$] is that something must be terribly wrong.

In the early 1980s the cosmological constant became quite unpopular with the introduction of new ideas, and the increased visibility of old ones, which are to be considered next. Thinking changed again at the turn of the century, when the observations were seen to require a positive value of Λ.

## 6.4 Inflation and Coincidences

Why is the universe close to the same everywhere? The exponential instability of the mass distribution in Einstein's static universe becomes a more slowly growing power law instability in an expanding universe, but that still means the close to uniform universe we see now would have to have grown out of very close to exact homogeneity in the very early stages of expansion. Can this initial condition be understood? As often happens, the question became deemed to be pressing when a promising answer turned up.

The central idea of what Alan Guth (1981) termed the inflationary universe, later inflation, is that a hypothetical close to constant and large energy density present in the early universe caused the universe to expand at an enormous rate. (In gravity physics the rate of expansion at escape speed is predicted to vary as the square root of the mass density. Large mass density means rapid expansion.) The enormous expansion during inflation was supposed to have stretched out the primeval, maybe quite chaotic, state of the universe; stretched by such a large factor that there are only tiny variations in the initial conditions across the minuscule bit of the primeval universe that we observe well after inflation. An implication of particular relevance for us is that the enormous expansion would have stretched out the mean radius of curvature of space sections to some enormous value. That would produce very close to flat space sections across the bit of the universe we can observe.

We can disregard debates about whether inflation really would present us with a close to homogeneous universe out of some sort of primeval chaos, and the complaint about the lack of a specific theoretical basis for the hypothetical enormous early mass density. Inflation has interesting features that we need not review; more important for our purpose is a lesson. The inflation scenario was welcomed as a truly elegant idea that produced the community assessment that space sections surely are flat, stretched out by inflation. This was a social construction, supported by little in the way of predictions that were not already being discussed for other reasons. There is nothing wrong with this; general relativity was a social construction with little empirical support in 1960. But the situation is to be borne in mind.

A commonly accepted opinion in the years around 1990 was that another argument also supports the conclusion that space sections must be flat. Suppose first the cosmological constant vanishes, and the universe is expanding at less than escape speed, so space sections are closed as in the surface of a sphere. Then there is a special time, an epoch in the course of evolution of this closed universe, when the expansion stops and the universe starts to collapse. The presence of a negative cosmological constant would contribute to the slowing of expansion, and again there would be the special epoch when expansion reverses to collapse. If there were a positive cosmological constant the special epoch could occur when the mass density had become small enough that the rate of expansion of the universe changed from being slowed by the attraction of gravity to being sped up by the effect of the cosmological constant. The special epoch in an open universe without a cosmological constant is the time at which the expanding universe escapes the pull of gravity. The situation is different in one particular case, where there is no cosmological constant and no space curvature. After the start of expansion, whatever that means, there is no special epoch in the course of evolution of this universe; the expansion at escape speed continues forever more.

Already in the 1960s it was reasonably clear that the mass density in our universe is in the range of values that would imply that the expansion of the universe could be approaching one of these special epochs, either just now escaping the pull of gravity, perhaps with the help of a positive cosmological constant, or else close to the time when expansion turns to collapse. In either of these cases we would have evolved to the point of taking an interest in the expanding universe at about the time of one of these special epochs. This would be a curious coincidence. Our existence requires a galaxy, which requires enough mass density for its gravitational assembly. But after assembly, general relativity predicts that whatever happens in our galaxy is quite independent of the evolution of the universe as a whole, apart from the possibility of another galaxy running into us, as would happen in the approach to a big crunch. So why the coincidence? None would be needed if there were no cosmological constant and our universe were expanding at escape speed, for in this case we would see expansion at escape speed whenever we happened to flourish. This universe, with no space curvature, no cosmological constant, and no curious coincidence, is known as the Einstein-de Sitter model.

Einstein and de Sitter (1932) offered a different argument for this model. In the relativistic theory, the present rate of expansion of the universe, measured by Hubble's constant, is determined by the sum of terms representing the mean mass density, space curvature, and the cosmological constant. The mass density is the only one of these three terms that we can be sure is present, so they argued that until we have better evidence simplicity suggests we set the other two terms to zero. One could go further and argue that, since matter is needed for our existence and space curvature and Λ are not, these last two terms surely would not be present in the maximally elegant universe. Einstein and de Sitter did not go that far. They remarked that

> The curvature is, however, essentially determinable, and an increase in the precision of the data derived from observation will enable us in the future to fix its sign and to determine its value.

Einstein had come to dislike the cosmological constant. Maybe that is why they wrote that a consistent cosmology "can be reached without the introduction of $\lambda$" (Einstein's original notation for Λ).

The arguments against the coincidence of epochs, and for the simplicity of the Einstein-de Sitter model, were well worth considering. Experience in physics has been that fundamental theory tends to simplicity, though this is difficult to frame in a falsifiable statement. I remember discussions of the coincidence argument against the presence of Λ and space curvature when I was a postdoc in Dicke's Gravity Research Group in the early 1960s. There are scattered comments about it in the literature, by Dicke (1970 p. 62), McCrea (1971, p. 151), and Dicke and Peebles (1979, pp. 506–507), but it was another of cosmology's "dirty little secrets."

Opinions about the Einstein-de Sitter model have evolved. Robertson (1955) commented that it is of "some passing interest." Bondi (1960), in the addendum to the second edition of *Cosmology*, remarked on "its outstanding simplicity." We have seen that Gunn and Tinsley (1975) expressed special caution in presenting evidence of possible detection of the effect of $\Lambda$. In the 1970s the coincidence argument persuaded me that the universe likely is Einstein-de Sitter.[4] I was led to change my thinking in the early 1980s by advances in measures that indicated the mean mass density is well below the Einstein-de Sitter value. The arguments are presented in Davis and Peebles (1983) and Peebles (1986).[5]

In the mid-1980s, and for the next decade, the argument from inflation for flat space sections, and the coincidence argument for that and against $\Lambda$, a term that is ugly anyway, led to the community feeling that the universe surely is Einstein-de Sitter. This was a social construction that many considered persuasive. But there was the evidence that the mass density is less than Einstein-de Sitter.

4. My thinking was reenforced by the results of statistical analyses of the galaxy space distribution, which showed that the galaxy two-point position correlation function is close to a simple power law on relatively small scales, with a break from the power law, first up and then more sharply down at a slightly larger scale, about where the spatial mass density fluctuations make the transition from large departures from homogeneity on smaller scales to small fractional departures on larger scales (Davis, Groth, and Peebles 1977, Fig.1; Soneira and Peebles 1978, Fig. 6). The break spoke to me of the transition from linear to nonlinear growth of clustering. The featureless power law on smaller scales suggested to me scale-invariant evolution, as in the Einstein-de Sitter model. But this argument is challenged by numerical simulations that indicate the galaxy position correlation function is close to a power law while the mass autocorrelation function is not, a curious situation.

5. In a little more detail, the Davis and Peebles (1983) analysis of the first reasonably fair sample of the distributions of galaxy angular positions and redshifts, obtained by Davis, Huchra, Latham, and Tonry (1982), indicated that the galaxy relative velocity dispersion at galaxy separations ranging from 0.2 Mpc to 6 Mpc is consistent with mean mass density about one fifth of the Einstein-de Sitter value. It was natural to object that this measure would miss mass that is more smoothly distributed than the galaxies, an effect known as biasing. The counterargument was that, if that were the case, then it would be curious that our measure of the density is close to the same over a considerable range of galaxy separations. Why were we not picking up the more smoothly distributed component at larger separations? I reviewed in Peebles (1986) this and other arguments that made the case for a low mass density look reasonably good to me, though not yet to the community.

Neta Bahcall led research that showed that the abundance and spatial clustering of positions of clusters of galaxies indicates the cosmic mean mass density is less than Einstein-de Sitter (Bahcall and Cen 1992), and demonstrated that the low mass density also is required to account for the relatively slow growth of the masses of the great clusters of galaxies (Bahcall, Fan, and Cen 1997). Ideas can be influential, however. I remember conferences where Neta Bahcall's evidence was not as well received as it ought to have been. I also remember a bright young colleague telling me, in effect, that I was only arguing for low mass density to annoy, because I knew it teases. I was serious, but I did know it teases. Such is the behavior of people.

Good evidence is even more influential than good ideas, of course. In the span of five years around the turn of the century, the two great programs to be described in Section 6.10 established a clear and convincing case that persuaded the community that the mean mass density is less than Einstein-de Sitter, but that space sections are flat[6] because there is a positive cosmological constant $\Lambda$. The value of $\Lambda$ is quite unreasonable compared to what is suggested by the sum of quantum zero-point energies, as we have seen, but $\Lambda$ nevertheless has been renamed dark energy (I believe by Huterer and Turner 1999).

We should pause to consider that in the years around 1990 there was little discussion of whether we were fooling ourselves by taking the relativistic theory so seriously. This is curious, because it is an enormous extrapolation of physical theory from its tests on the scale of the solar system and smaller. The excuse was that we had no alternative to the general theory of relativity, and that it offered a plausible framework for predictions to compare to the broad variety of tests that were being considered. That is a serious practical consideration, but of course not a real excuse. We were lucky; the consistency of predictions and observations to be discussed now makes a persuasive case that general relativity with Einstein's cosmological principle is a useful approximation.

---

6. Recall that, given Hubble's constant and the mean mass density, the relativistic theory fixes the sum of the terms representing the effects of the cosmological constant and the curvature of space sections. This is expressed in the Friedman-Lemaître equation in the footnote on page 176. If space sections are flat and the mass density is low then the theory requires a positive value of $\Lambda$.

The detection of the effect of Einstein's cosmological constant means that we flourish at a special epoch, just as the rate of expansion of the universe is changing from slowing due to the attraction of gravity to speeding up due to the effect of the cosmological constant. This violates the coincidences argument. The evidence of flat space sections agrees with the stretching argument from inflation. The empirical case for flat space sections is celebrated for its support for inflation. I have not heard expressions of regret for the loss of the coincidences argument against $\Lambda$. You win some, you lose some.

## 6.5 Baryonic and Subluminal Matter

Protons and neutrons, the particles we are made of, and the other strongly interacting particles, are said to be baryons. In cosmology baryonic matter includes these particles with their electrons. Until the 1970s it was considered natural to take it that all matter is baryonic, but long before that the Swiss physicist Fritz Zwicky had found the first hint that there may be more to it.

Some galaxies are in great clusters. In line with the galaxies of stars, one is tempted to say that these clusters are galaxies of galaxies. The Coma Cluster (named for its position in the constellation Coma Berenices) is the nearest really large one. Zwicky (1933) assembled measurements of spectra of eight galaxies in this cluster. The wavelength differences among the spectra, interpreted as Doppler shifts due to their relative motions, indicated that these galaxies are moving relative to each other so rapidly that the gravitational attraction required to hold them together requires considerably more mass than the sum of the masses of the galaxies in the cluster. It was commonly said that the mass required to hold this cluster together is missing. This could mean the cluster is flying flying apart, but Zwicky (1937) showed that the compact and regular central distribution of galaxies in this cluster certainly does not suggest that. Zwicky (1933) pointed out that the mass of the cluster may be largely "dunkle Materie." Indeed, to have escaped detection the matter cannot be very luminous. To avoid confusion with the idea of the nonbaryonic dark matter considered in Section 6.6, let us translate Zwicky's term to subluminal matter.

Some remembered Zwicky's phenomenon, but few discussed it in print until the 1990s, after the idea of nonbaryonic matter had become popular. Before then it was just another of the "dirty little secrets" of physical cosmology.

Until quite recently there were even fewer citations of Horace Babcock's paper on measurements of velocities of four clouds of baryonic plasma in the outer parts of the nearest large spiral galaxy outside our Milky Way. This is the Andromeda Nebula, also known as M 31. Babcock observed places in the galaxy where hot young stars have ionized the interstellar matter around them. The recombination radiation from this plasma produces emission lines in the spectrum; hence the name emission line regions (later named H II regions, meaning ionized hydrogen, while H I regions contain largely neutral atoms). These emission lines are good targets for measurements of Doppler shifts.

Babcock discovered that the velocities of these regions relative to the mean for the galaxy required an unexpectedly large mass in the outer parts of the galaxy to produce the gravitational attraction to hold these regions in the galaxy. Babcock referred to Zwicky's discussion of measurements of masses of galaxies, but not to Zwicky's (1933, 1937) subluminal matter in the Coma Cluster, or to Smith's (1936) similar evidence for subluminal matter in the closer smaller Virgo Cluster of galaxies. Babcock remarked on dark features in nebulae, the galaxies, that might be caused by absorption of starlight, and he suggested that this absorption may hide a considerable mass in stars in the outer parts of M 31. There is no subsequent evidence of this.

Another possibility Babcock mentioned is

> that new dynamical considerations are required, which will permit of a smaller relative mass in the outer parts.

Without making too much of what Babcock may have meant by this we should note that the Newtonian gravity physics that is successfully applied to the solar system is extrapolated by some ten orders of magnitude to the length scales of Babcock's observations. This great extrapolation could have been more commonly questioned; perhaps Babcock had something like this in mind when he mentioned new dynamical considerations. The extrapolation has been found to be accurate, however, and Babcock's large velocities point to the presence of subluminal matter in the outer parts of the spiral galaxy M 31.

Astronomers speak of the rotation curve of a spiral galaxy. It is the speed of circular motion of stars and gas in the disk as a function of distance from the center of the galaxy. Close to the center of a typical spiral galaxy the rotation curve is observed to increase with increasing distance from the center about as expected from the gravitational pull of the mass in the seen stars that is required to hold the matter in circular motion. But in the outer faint parts the rotation curve tends to be flat, that is, close to independent of distance from the center. If the mass were where the stars are, then in the outer parts of the galaxy the rotation curve would decrease with increasing distance from the center. The flat curve indicates the presence of subluminal matter whose gravity holds the stars and plasma in circular motion even in the outer parts of the galaxy. Babcock's (1939) probes of the rotation curve of M 31 gave the first hint of subluminal matter around galaxies, of course assuming we have an adequate theory of gravity. Babcock's discovery was not widely advertised until after the revolutionary advances in the cosmological tests at the turn of the century.

The history of additions to the evidence of flat rotation curves and the presence of subluminal matter in the outer parts of the conveniently close spiral M 31 illustrates how advances in technology drive and are driven by pure research. Walter Baade (1939) reported that

> Although a large number of orthochromatic and red-sensitive plates have long been on the market, their sensitivities in the yellow and red are so low compared with the blue sensitivity of ordinary plates that prohibitive exposures are necessary to reach really faint objects. Last fall Dr. Mees of the Eastman Company sent us for trial a new red-sensitive emulsion, labelled H-$\alpha$ Special, which proved to be so fast in the red that one may say without exaggeration that it opens up new fields in direct astronomical photography.

Hydrogen is the most abundant element in interstellar matter. The recombination of ionized to atomic hydrogen produces the prominent H-$\alpha$ (Balmer series) line in the red (at wavelength

656 nm = 6560 Å). Baade could identify emission line regions by this red line because it causes these regions to appear in images of the galaxy on the new red-sensitive plates and not on the blue-sensitive ones. He catalogued 688 of these regions in the Andromeda Nebula.

Nicholas Mayall (1951) measured the redshifts of a sample of Baade's emission line regions in M 31. His conclusion is that

> At nuclear distances greater than 65' to 70', to nearly 100' north-following and 115' south-preceding, there is good evidence of a decrease in rotational velocity with increasing nuclear distance. In other words, the "turn-around" points apparently occur at nuclear distances of 65' to 70', where the main body of the spiral as shown on most ordinary photographs appears to end.

To the modern eye the outer parts of Mayall's rotation curve look flat. Mayall would have been conditioned to think that the mass of the galaxy ends where the starlight ends, and that past that "turn-around" point the rotation curve decreases with increasing distance. That is what he saw. We can take the difference between what Mayall saw and what is seen now to be an example of Kuhn's (1970) paradigm shift of normal science.

The 25-metre radio telescope at Dwingeloo, the Netherlands, another advance in technology, was large enough that it had the angular resolution to probe the redshift of the 21-cm line from atomic hydrogen across the face of M 31. Van de Hulst, Raimond, and van Woerden (1957) reported measurements of the rotation curve that are in line with and more accurate than Mayall's (1951) optical measurements. They did not mention a turn-around point. Maartin Schmidt (1957) used these data to estimate the mass distribution in this galaxy; he concluded that the interpretation required a better measurement of the distribution of starlight. At his suggestion, Gérard de Vaucouleurs (1958) measured the surface brightness across this galaxy and compared it to the mass distribution derived from the redshift measurements. He concluded that the ratio of mass to starlight increases with increasing distance from the center, as if the mass in the outer parts of this galaxy were dominated by subluminal matter.

In an early step from photographic plates to far more sensitive digital detectors, the Carnegie Institution of Washington had encouraged Kent Ford, at the east coast branch of the Institution, to look into image tube intensifiers made by RCA in Lancaster, Pensylvannia. They would improve efficiency of detection over the use of photographic plates alone. His colleague, Vera Rubin, led the use of the image tube spectrograph to get greatly improved measurements of spectra of emission line regions in Baade's finding list for M 31. Rubin (2011) recalls learning about Baade's list: "For us, this remarkable gift of M31 stars and emission regions was truly coming from heaven." It allowed them to guide on offsets of the emission regions they could not see from stars they could see. In another important technical advance Rubin and Ford (1970a) reported that

> the use of image intensifiers has made it possible to reduce observing times by a factor of 10, making it possible to complete fairly ambitious observing programs in a few observing seasons.

They obtained a still tighter measurement of the rotation curve of M 31, clearly showing its flat outer part.

Roberts and Whitehurst (1975) used the 300-foot (90-m) telescope in West Virginia, with its considerable improvement in angular resolution over the pioneering 25-m dish, to extend the measurements of the 21-cm line rotation curve to larger distances on one side of the galaxy. (Measurements on the other side are more difficult because the Doppler-shifted wavelengths of the 21-cm lines from M 31 and from the Milky Way in the direction of M 31 are similar.) The rotation curve remains flat at this greater distance from the center of the galaxy.

Meanwhile, Kenneth Freeman (1970) and Seth Shostak (1972) used 21-cm observations to show evidence of subluminal matter in the outskirts of two other spirals, NGC 300 and NGC 2403. Freeman does not seem to have been aware of the rotation curve measurements for M 31, or of the evidence of subluminal matter in clusters of galaxies. Shostak referred to the prior evidence for subluminal matter around spiral galaxies from Freeman and from Rubin and Ford.

At the same time, Hohl (1970), Ostriker and Peebles (1973), and others were demonstrating that a rotationally supported disk such as a spiral galaxy with a roughly flat rotation curve is unstable if the gravitating mass that holds it together is all in the disk and rotating with the disk. The instability is remedied by placing most of the mass in a more slowly rotating and stable halo of subluminal matter. This consideration added to the evidence that the mass in the outer parts of spiral galaxies is subluminal.

The emergence of evidence of subluminal matter is not well described as a discovery. It was a growing realization, from the indication of subluminal matter in clusters of galaxies in 1933, to the first hint of subluminal matter around galaxies in 1939, to an excellent case by 1975. But it is fair to say that the evidence for subluminal matter around galaxies presented in 1970 by Rubin and Ford, and independently by Freeman, with the evidence from the instability of massive disks, from Hohl, and Ostriker and Peebles, qualifies as a Merton multiple recognition that the mass in a galaxy does not seem to be where the starlight is.

The presence of subluminal matter is interesting, but from the evidence in the 1970s it need not have been revolutionary. Roberts and Whitehurst (1975) pointed out that

> Dwarf M stars, the most common type of star in the solar vicinity, of number density of a few tenths $pc^{-3}$ are adequate to explain the required mass.

Let us consider now the path to a revolutionary idea, that the subluminal matter is not baryonic.

## 6.6  Dark Matter

In the 1970s two sets of Merton multiples introduced the ideas that the subluminal matter around galaxies and in clusters of galaxies is one or another kind of neutrino with a nonzero rest mass, and that there might be a cosmologically interesting mass density in this matter that is not baryonic, not the familiar protons and neutrons. It led to the idea of the dark matter of the established cosmology. The nature of the dark matter is not known at the time of writing. There are many ideas, but the thought of a neutrino with a significant rest mass got us here. Maybe neutrinos are the answer.

At the start of the 1970s it was known that there are at least two types of neutrinos with their partner leptons: electrons and muons. Evidence of the tau lepton and its neutrino was found toward the end of the decade. These neutrinos would have been produced and annihilated by thermally driven interactions with electron-positron pairs in the sea of radiation in the early hot stages of expansion of the universe. That means the present number density of these fossil neutrinos is determined by the measured temperature of the fossil thermal radiation (Sec. 5.4). It works out to a neutrino number density of about $100$ cm$^{-3}$ in the present universe. If the electron or muon neutrino had a rest mass of a few tens of electron Volts (where the energy of one electron Volt amounts to a mass of about $10^{-33}$ g), then the product of the number density with the neutrino rest mass would be a mass density comparable to or larger than the baryonic mass density. This certainly would be interesting for cosmology, and it is the first model for nonbaryonic weakly interacting dark matter.

This neutrino model was introduced by Cowsik and McClelland (1972) in the US, and by Marx and Szalay (1972) in Hungary, independently, a Merton doubleton. Gershtein and Zel'dovich (1966) had pointed out that a reasonable limit on the mass density of a relativistic cosmological model, from its expansion rate and expansion time, with an estimate of the number density of thermal neutrinos set by the temperature of the recently discovered fossil radiation, implies an upper bound on the muon neutrino rest mass. Since the laboratory measurements of the rest mass were not yet very constraining, this was an interesting result, but not recognition of a possible connection to the phenomenon of subluminal matter. Marx and Szalay knew the Gershtein and Zel'dovich constraint; Cowsik and McClelland rediscovered it. Both knew Zwicky's evidence of subluminal matter in the Coma Cluster, and recognized that it could be neutrinos remnant from the hot big bang, if neutrinos had the proper non-zero rest mass. They do not seem to have noticed the more extensive but not yet well advertised evidence of the presence of subluminal matter in massive halos around galaxies.

In a follow-up paper on the structure of the Coma Cluster, under the assumption that its mass is largely the neutrino dark matter, Cowsik and McClelland (1973) wrote that

> Though the idea is undoubtedly not new, it does not appear to have been presented in published form before.

As it happened, Marx and Szalay published the idea at essentially the same time, independently. The common occurrence of Merton multiples suggests that the massive neutrino idea might have occurred to still others, as Cowsik and McClelland suggested, though no one has come forward.

The idea that neutrinos with rest mass of a few tens of electron Volts could be the subluminal matter, and could make an important contribution to the cosmic mean mass density, certainly was interesting. Neutrinos were known to exist. They interact only weakly with baryonic matter, making it easy to see why this kind of subluminal matter would have been detected only by its gravity. Interest in the idea was increased by a report of laboratory detection of the electron neutrino rest mass, at a few tens of electron Volts, about what would be wanted for this kind of subluminal matter. The evidence came from the decay of tritium, the unstable heavy isotope of hydrogen, to the lighter isotope of helium, with the emission of an electron and an antineutrino. These two particles share the decay energy. If the electron antineutrino had a nonzero rest mass it would reduce the maximum energy the electron can carry away, because the antineutrino would have to be given at least the energy equivalent of the rest mass. The reported detection of this effect on the electron decay energy certainly was influential, but it was found to be wrong. The known neutrinos do have nonzero rest masses, but they are much smaller than contemplated in the 1970s.

In 1977, ideas circulating in the particle physics community motivated the proposal of another candidate for nonbaryonic subluminal matter, a new neutrino to be added to the known electron, muon, and tau types. The new neutrino would have a much larger rest mass, about $3 \times 10^9$ electron Volts, or 3 GeV. The relatively large rest mass means most of this kind of neutrino would have annihilated as they were being thermally produced in the early universe. But if these neutrinos had the same weak interaction as the known families, which fixes the rates of their creation and annihilation, then at this rest mass the number density of the small fraction of the neutrinos that escaped annihilation, multiplied by the 3 GeV mass per neutrino, would amount to a mass density that would be interesting for cosmology.

The idea of this second kind of nonbaryonic matter was proposed in five papers, submitted for publication in the space of two months. In increasing order of date of reception by the journal, they are Hut (1977); Lee and Weinberg (1977); Sato and Kobayashi (1977); Dicus, Kolb, and Teplitz (1977); and Vysotskii, Dolgov, and Zel'dovich (1977). The proposals were independent, as far as I can determine, making it a remarkable Merton quintuple.

These authors had a common interest in particle physics. At the time, evidence was emerging for the existence of the third kind of neutrino. Its partner, the tau lepton, has a much larger rest mass than the muon, which in turn is much more massive than the electron. The evidence was that the tau mass is about 2 GeV, and that the associated neutrino rest mass could be as large as 0.6 GeV (Perl, Feldman, Abrams et al. 1977). The bound on this tau neutrino mass is much tighter now, but at the time the more generous bound could have encouraged the thought of a fourth lepton family with a still larger neutrino rest mass, perhaps 3 GeV. All five of the 1977 papers have citations to one or more of the papers by Szalay and Marx, or Cowsik and McClelland, on the idea that one of the known types of neutrinos has a much smaller but cosmologically interesting rest mass. The authors of the five papers may have been aware of the experimental evidence that the electron-type neutrino has a rest mass of a few tens of electron Volts, or of the refutation of this evidence. All five knew Zwicky's evidence of subluminal mass in the Coma Cluster of galaxies, but none showed recognition of what had become an array of evidence of subluminal matter around large spiral galaxies. Apart from these hints the occurrence of this quintuple leaves us with thoughts of "something in the air."

The first of the ideas about nonbaryonic subluminal matter, neutrinos with rest mass around 30 eV, was seriously challenged by the conditions for formation of cosmic structure: galaxies and their clustered spatial distribution. These neutrinos would have been moving about at close to the speed of light in the early universe. That led to the later name, hot dark matter, or HDM. The rapid motions would have smoothed the primeval departures of the mass distribution from exact homogeneity, ironing it out to a mass characteristic of clusters of galaxies. It would mean that, if the first generation of mass concentrations grew by gravity, then proto-clusters would have formed first, and would have to have fragmented to produce the galaxies. This scenario was explored largely by Zel'dovich's group, but others took an interest in it. It has a serious problem, however. Young proto-clusters would have to have scattered the fragments, young galaxies, broadly, because most galaxies are not near clusters. But gravity does not scatter, it gathers together. In this picture most galaxies would be in or near clusters. This was known to be quite contrary to the observations.

The much more massive fourth neutrino introduced in 1977 would have moved about much more slowly in the early universe. For this reason it was later named cold dark matter, or CDM. The negligible smoothing of the primeval mass distribution in this picture meant that galaxies could have been assembled by gravity before gathering into groups and clusters of galaxies. This looked much more promising. The idea that this hypothetical fourth, quite massive, neutrino is the dark matter of the established cosmology is still viable, provided that the rest mass and the strength of inter- action with baryonic matter are adjusted to account for the lack of dark matter detection in laboratory searches. There are other ideas about the nature of the dark matter, but these massive neutrinos gave us the prototype for the cold dark matter, CDM, cosmology introduced five years later.

The idea in 1977 that the subluminal matter may be the mas- sive neutrinos in a fourth family caught the attention of people who knew the astronomical evidence: Gunn, Lee, Lerche et al. (1978) and Steigman, Sarazin, Quintana, and Faulkner (1978). They pointed out that, if the subluminal matter is CDM (though it did not have the name yet), then the formation of a galaxy in the expanding universe would be expected to have started with the gravitational assembly of a diffuse cloud of the cosmic mixture of baryonic and nonbaryonic cold dark matter. The baryonic matter would settle further, by radiative dissipation of energy, to a more compact central concentration, a young galaxy. This concentration would fragment in turn, again by radiative dissipation, to produce stars. The scenario seemed to offer a promising approximation to what was known about large spiral galaxies and their subluminal halos. The point is important, but at the time it did not add much to the case for nonbaryonic matter, because White and Rees (1978) arrived at essentially the same picture, independently, by postulating that subluminal matter is the remnant of an early generation of stars that have faded away and are now subluminal. The similarity of the two pictures makes it a Merton doubleton. The difference—baryonic or nonbaryonic—is important but was not much debated because community attention soon was captured by attractive features of nonbaryonic cold dark matter, as follows.

## 6.7  The CDM Theory

The change of thinking that drove the general acceptance of non-baryonic cold dark matter, CDM, grew out of the considerable difference between the quite smooth fossil sea of thermal radiation and the prominent concentrations of matter in galaxies and groups and clusters of galaxies. If gravity drew matter together to form these mass concentrations, how could this process have avoided a considerable disturbance to the radiation?

A trial step to an answer was motivated by measurements in the early 1980s that yielded tentative evidence that the fossil radiation temperature varies across the sky by a few parts in ten thousand ($\delta T/T \sim 10^{-4}$). I had to pay attention to this because it was proposed, tentatively but independently, by two groups, Fabbri, Guidi, Melchiorri, and Natale (1980) and Boughn, Cheng, and Wilkinson (1981). Boughn et al. were colleagues at Princeton University, in the Gravity Research Group. I proposed a model (in Peebles 1981) to reconcile this anisotropy with the clumpy distribution of the mass in galaxies. It was contrived[7] to fit the purpose, but arguably not unreasonable. When Fixsen, Cheng, and Wilkinson (1983) withdrew the earlier claim of detection, and showed that the

---

7. The main trick was a substantial tilt of the initial conditions to favor large primeval departures from homogeneity on relatively small scales, to encourage galaxy formation, and small departures on the larger scales being probed by most of the fossil radiation measurements. I don't regret the detour; it is a normal and healthy part of the game.

sea of radiation is even smoother than allowed by my model, I came up with another one (Peebles 1982). That was before announcement of the new Fixsen et al. measurement, but I knew about it; we were colleagues.

When I introduced this second cosmological model, I had been spending a lot of time on the fascinating (to me and like-minded folks) development and application of statistical measures of the spatial and velocity distributions of the galaxies. Many of the results are collected in Groth and Peebles (1977) and Davis and Peebles (1983), with a lot more in Peebles (1980). I started this project with the vague thought that it might reveal statistical patterns that hint to how the distribution of the galaxies got to be the way it is. The results (Davis, Groth, and Peebles 1977; Peebles 1984a) seemed to me to be what one would expect if the galaxies and their clumpy space distribution had grown by gravity out of small primeval departures from an exactly homogeneous early universe. The great difference between the smooth distribution of the fossil radiation and the clumpy distribution of the matter seemed to challenge my thinking about this. It led me to devise the new cosmology (in Peebles 1982) that became known as the CDM model. I meant it to be a counterexample to the argument that the idea of gravitational growth of cosmic structure was in trouble.

My proposal started with the assumption that most matter is nonbaryonic. It would be dark, and it would at most weakly interact with baryonic matter and radiation. This would allow the sea of thermal radiation to slip freely through the nonbaryonic matter as gravity gathered matter to form the mass concentrations characteristic of galaxies. The radiation would be disturbed by the drag of the baryonic plasma, but most of the mass was supposed to be nonbaryonic. The radiation would be disturbed by the gravity of the growing mass concentrations. That effect can be computed in the general theory of relativity.

I adopted the scale-invariant initial conditions for the growth of cosmic structure discussed in Section 6.1, again for simplicity. It requires the choice of just one parameter, the size of the spacetime curvature fluctuations produced by the departures from exact homogeneity. The primeval space distributions of the two kinds of matter and of the primeval radiation could differ, but again simplicity dictated that all three are the same. (This is known as adiabatic initial conditions.) I took it that the galaxies are fair tracers of the mass distribution. As I have mentioned, we had well-checked statistical measures of how the galaxies are distributed and moving, and I applied these measures to the mass. Finally, I adopted the simplest of all reasonable-looking relativistic cosmologies, the Einstein-de Sitter model discussed in Section 6.4. This is the model with no cosmological constant and no space curvature. I mentioned that my statistical analyses had yielded what I considered a reasonably good case that the mass density is less than Einstein-de Sitter, but taking account of that would require introducing another parameter, space curvature or the cosmological constant, a complication to be avoided in a counter-example.

The elements of the theory for the computation of the predicted gravitational disturbance of the fossil radiation caused by the gravitational growth of cosmic structure were already in place, from Sachs and Wolfe (1967) and Peebles and Yu (1970). The great challenge was to the experimentalists: develop the methods capable of detecting the predicted statistical patterns of departures from an exactly uniform sea of radiation, which was expected to be at the level of a few parts in a million. To my surprise the effect was detected, and the measurements have become far more accurate than that.

Smoot, Bennett, Kogut et al. (1992) announced the first detection of the large-scale variation of the fossil radiation temperature across the sky. Their instrument was on the COBE satellite, along with the one discussed on page 130 that demonstrated the thermal spectrum of the radiation. Within all the uncertainties, my 1982 prediction agrees with these measurements made a decade later. There were several reasons. I had been establishing statistical measures of the galaxy distribution that I could rely on. I assumed the mass is distributed with the galaxies, which turns out to be a reasonable approximation. The mass determines the gravity that disturbed the radiation, which can be computed. And my other assumptions in the CDM model proved to be about right, a combination of good luck and good management guided by what I could see might work.

Turner, Wilczek, and Zee (1983) independently introduced elements of this CDM model. They were interested in the possible role of particle physics in cosmology, not the contrast between the smooth distribution of the fossil radiation and the clumpy distribution of the matter. Their less detailed computation of the effect of gravity on the fossil radiation led them to conclude that the model is not promising.

I saw nothing wrong with the CDM model as an example of how gravity could assemble cosmic structure while leaving a much smoother sea of radiation, but I could think of other models that would do that, so I was a little disconcerted by the enthusiasm with which the CDM model was received. There was the problem mentioned in Section 6.4, that the straightforward reading of the evidence was that the mean mass density is less than in the Einstein-de Sitter model I had used. The evidence for lower mass density looked good to me, so in a second paper on the CDM idea (Peebles 1984b) I discussed the advantage of lowering the mass density, to agree with the evidence, and adding the cosmological constant Λ, to keep space sections flat, to agree with inflation. This became known as the ΛCDM theory.

## 6.8 The ΛCDM Theory

In the original CDM and then ΛCDM models the initial conditions for cosmic structure formation are scale-invariant. Later precision measurements of the pattern of the fossil radiation distribution require adjusting the initial conditions to favor slightly larger space-time curvature fluctuations on larger length scales. This means the degree of the departure from scale-invariance, what became known as the tilt, becomes a free parameter to be adjusted to fit the measurements.[8] This brings the theory to a nominal seven free parameters. We can take them to be

---

8. Simple implementations of the cosmological inflation picture discussed in Section 6.4 predict a good approximation to the scale-invariant initial conditions discussed in Section 6.1. A modest and arguably natural adjustment of the implementation tilts initial conditions to favor somewhat larger spacetime curvature fluctuations on larger length scales. As it happens, the fit to the observations requires this slight tilt. There would be trouble if this postulated tilt extended to arbitrarily large length scales. But that dangerous scale is so large it usually is ignored.

1. The present mass density in baryonic matter.
2. The present mass density in nonbaryonic cold dark matter.
3. The value of Einstein's cosmological constant.
4. The mean curvature of space sections.
5. The amplitude of the primeval spacetime curvature fluctuations.[9]
6. The degree of tilt of the spectrum of the curvature fluctuations.
7. The optical depth for scattering of the fossil radiation by free intergalactic electrons.[10]

Given these quantities, general relativity fixes the value of Hubble's constant.

The usually unspoken assumption is that we can apply Einstein's general theory of relativity. By the 1990s general relativity had been shown to pass demanding tests from the timing of light pulses within the solar system. It was natural to apply this theory to the large-scale structure of the universe, but it was an assumption, in effect a parameter choice. And it is a spectacular extrapolation from measurements in the solar system, which span lengths up to about $10^{13}$ cm, to the length scales of cosmology, roughly $10^{28}$ cm, fifteen orders of magnitude larger. No evidence apart from the indication of subluminal matter argued against it, but very little empirical evidence supported it prior to the advances of the cosmological tests to be discussed.

Let us consider also a lesson from sociologists, who observe the tendency to circular constructions in natural science as in all human activity. In the 1980s, when I introduced CDM and the first generation of the ΛCDM model, I saw that the galaxy distribution and motions look as though they had grown by gravity, so that is what I assumed. This means it is not entirely surprising that the theory offers a reasonable fit to the measured distributions

9. As a technical aside, let us note that the curvature fluctuations are taken to be a random Gaussian process that is defined by its mean and standard deviation. This is a close approximation to what simple implementations of cosmological inflation predict, and it agrees with the close to Gaussian nature of the measured fluctuations of the fossil radiation.

10. Ionizing radiation from sources not yet convincingly identified ionized the diffuse intergalactic baryons at redshift around z ~ 10. It freed electrons that scatter the fossil radiation. This slightly smooths the pattern in its distribution across the sky. About five percent of the fossil radiation has been scattered (Planck Collaboration 2020).

and motions of the galaxies. In effect, my assumption of the gravitational growth of cosmic structure was a free parameter that I chose to agree with what I recognized in the data. But this circularity was limited; my impression was a qualitative judgment based on sparse data. The tests of ΛCDM to be discussed here and in Section 6.10 are results of quantitative analyses of abundant precision measurements that check reliably computed predictions This is a big difference.

It is essential to be able to compare predictions to reliable measurements, of course, but the measurements are not so useful if the predictions are not reliable. Thus it is important that key ΛCDM predictions are reliably computed, in perturbation theory. This includes the successful predictions of the statistical properties of the distribution of the fossil radiation across all angular scales that have been measured so far, and the successful prediction of the statistical properties of the galaxy spatial distribution on a considerable range of relatively large scales. The situation is different on the smaller length scales of galaxies because galaxy formation is a complex process, to be modeled with the introduction of free parameters to approximate how stars formed and disturbed what gravity had piled up in young galaxies. This means the successes of simulations of galaxy formation argue for ΛCDM, but not very strongly. The theory and observation of the distribution of the fossil radiation and the large-scale distribution of the galaxies are more secure, and so far they are consistent with the measurements.

With due attention to these remarks, I conclude that ΛCDM successfully brings theory into agreement with far more abundant observations than there are adjustable parameters, the seven explicit ones and the more difficult to count implicit ones. These tests are reviewed in Section 6.10. But first let us consider the situation in the years around 1990, when the community was divided on which adjustments of the CDM model might yield a good approximation to what we are taking to be reality.

## 6.9 Confusion

In the years around 1990 the search for a reasonably interesting cosmology was enlivened by the accumulation of pieces of evidence that seemed instructive but could be interpreted in different ways. Despite the resulting confusion there were several reasons why few questioned the hypothesis of nonbaryonic matter.

To begin, Gamow's (1948a,b) idea about element formation in a hot big bang (discussed in Sec. 5.3) is based on well-tested nuclear physics, and the fit of theory to observations of the abundances of the light element isotopes of hydrogen and helium looked satisfactory. But the fit required that the mass density in baryons be about 5% of the density in the Einstein-de Sitter model that was the community favorite. This baryon density is even smaller than most dynamical measurements of the total, which scattered around 20 or 30 percent of the Einstein-de Sitter value. Nonbaryonic dark matter could make up the difference, because it would not take part in the nuclear reactions in Gamow's theory. There is the assumption that nuclear physics has not changed since the time when the temperature of the fossil radiation was some ten orders of magnitude larger than it is now. This is a considerable extrapolation back in time, to be checked by the degree of consistency of all the apparently relevant information, the usual goal.

The evidence of the presence of subluminal matter around galaxies seemed to be clear, and it was reasonable to consider the idea that it is nonbaryonic rather than low mass stars. A psychological factor on the nonbaryonic side was that this picture offered an easy entry to exploration of how the galaxies formed and developed their observed pattern of clustering. Recall that the hypothetical nonbaryonic CDM is just a gas of particles with an exceeding long mean free path. Their behavior is relatively easy to follow in numerical simulations. Baryonic matter in galaxies displays the complexities of shock waves and turbulence, along with the complications attendant to the formation of stars and the effects of their winds and explosions. This means the behavior of a concentration of baryons with the mass and size of a galaxy is complex. If most of the mass were the much simpler CDM, then an easy start to analyses of the formation of galaxies and their clustered space distribution would be to ignore the baryons at first and follow the gravitational formation of concentrations of dark matter. The results in the 1980s looked promising. The natural next step was to introduce models for the behavior of the baryons in increasing detail, guided by what was found to produce reasonable-looking approximations to galaxies. It helped keep people interested in nonbaryonic matter.

Simplicity does not add to the evidence for CDM, of course, but it is a good reason to explore the assumption. This way of thinking has directed research to productive directions, on occasion. Maxwell's decision to put aside the search for a mechanical model of the ether and go forward with the formulation of field equations gave him a much simpler task that led to a successful theory. The simplicity of nonbaryonic CDM earned it a prominent place in the debates in the years around 1990 on the nature of a physical theory of cosmology that would pass the tests that were deemed relevant. There was no guarantee that this simplicity would make CDM a useful direction to explore, but we were lucky.

A feature I disliked about my 1982 CDM model was the large mass density, which I considered contrary to a straightforward interpretation of significant evidence. My 1984 version, ΛCDM, remedied that, but others disliked the cosmological constant. Measurements of the angular distribution of the fossil radiation were important in resolving these issues, but until the turn of the century preliminary assessments of these measurements tended to be confusing. We did have reliable statistical measures of the patterns in the distribution and motion of the galaxies, but there were doubts about how well the galaxies trace the mass, because if that were so it would mean the mass density is less than the favored Einstein-de Sitter value. Added to this were the occasional misunderstandings of the properties of galaxies and clusters of galaxies. The result was a confusion of thoughts about how to improve the original CDM model. Let us pass over the details entered in *Cosmology's Century* (Peebles 2020), and instead contemplate the variety of ideas, without attribution. We are interested in the big picture.

In the warm dark matter model, WDM, random initially large motions of the dark matter particles set a smoothing scale that would define a lower bound on the masses of the first generation of mass concentrations. This minimum mass might be observationally interesting, maybe a mass characteristic of the galaxies. The idea grew out of thinking in particle physics, but the value of a primeval minimum mass became a free parameter as thinking in particle physics evolved in other directions.

The tilted cold dark matter model, TCDM, assumes the mass density—baryonic plus nonbaryonic—is the Einstein-de Sitter value, thus eliminating the need for the cosmological constant to keep space sections flat. With the other assumptions of CDM, the fit to the measurements of the large-scale anisotropy of the fossil radiation would predict unreasonably massive clusters of galaxies. To get around that, the model assumes a considerable tilt of the primeval mass fluctuations from the scale-invariance assumed in CDM, in the manner discussed in Section 6.1. The assumption of near scale-invariance does follow from cosmological inflation in a reasonably natural way, but one can argue that the tilt in TCDM also is natural in inflation, with allowable adjustments of the scenario. This model does not address the problem of why the more direct dynamical measurements make a reasonable case that the mass density is less than Einstein-de Sitter, but a common feeling was that the dynamical mass measurements surely are systematically biased low.

In the mixed dark matter model, MDM, also known as cold + hot dark matter, or CHDM, the nonbaryonic dark matter is a mix of CDM and the HDM of the original idea of neutrinos with rest mass of a few tens of electron Volts. The initially cold component would be the subluminal matter concentrated around galaxies and in clusters of galaxies. The hot component would be more broadly spread out. Observations on relatively small scales thus might give the impression that the CDM is all the dark matter there is, missing the HDM. Maybe the total mass density, HDM added to CDM, is the Einstein-de Sitter value. The DDM (for decaying) picture would do much the same, by having some of the dark matter decay into rapidly moving dark particles that would serve as a hot component, leaving some cold dark matter to cluster around galaxies. Again, maybe there would be no need for the cosmological constant.

The $\Lambda$CDM model I introduced in 1984 accepts the evidence for low mass density and adds the cosmological constant to keep space sections flat. The $\tau$CDM model adds parameters to allow exploration of modest adjustments of $\Lambda$CDM.

Exploration of these models in the 1990s was healthy; one of the ideas might hit on the right direction. It is notable, however, that discussions seldom ventured from the idea that most matter is nonbaryonic. There was the different path taken by Ostriker and Cowie (1981). They started with the observation that the energy released by nuclear burning and explosions in stars is observed to rearrange baryons, as in the winds of plasma flowing out of star-forming galaxies. Maybe explosions that were even more energetic piled up matter to form the galaxies. The effect of the plasma on the distribution of the radiation was a complicated issue, but this proposal was published just before nonbaryonic matter was added to cosmological models. The explosion idea certainly was worth considering, but it was challenged by the observed small relative motions of neighboring galaxies and their large streaming motions. This is a readily visualizable result of the gravitational assembly of galaxies, but not explosions.

I introduced the PIB model (for primeval isocurvature baryon) in Peebles (1987a,b), for the purpose of a counterexample to show that we did not need this hypothetical nonbaryonic matter I had introduced to cosmology, provided you were willing to live with an awkwardly low mean mass density of the baryonic matter and the rather contrived initial conditions. The model assumes the baryonic matter, which is all the matter, was already in a clumpy state in the early universe, and that the radiation was distributed so as to keep the total mass uniform.[11] The PIB model was viable well into the 1990s, when it was falsified by the improving measurements of the anisotropy of the fossil radiation on angular scales of about one degree, where CDM-like models predict the observed bump in the anisotropy and PIB wrongly predicts a dip.

A nonspecialist looking into research in cosmology in the 1990s might have been dismayed by the many different ideas, all arguably promising, for the theory of our one universe. The situation could have been an entry in the book mentioned on page 58, *The Golems* (Collins and Pinch 1993). Their examples of aberrant thinking in natural science, real or apparent, are part of the normal course of research. The exchanges of ideas about cosmology were healthy,

---

11. The mass density of the radiation is smaller than the matter density now, but it would have been much larger in the early universe because expansion lowers the mass density in radiation faster than for matter. This means there would be no immediate problem with the assumption of a clumpy primeval baryon distribution compensated by small deviations from a smooth sea of radiation.

apart from the reluctance to explore more adventurous ideas. Might broader speculations in the 1990s have produced other viable cosmological models without nonbaryonic matter? We cannot know, because two great advances in observational programs in progress at the turn of the century made the question uninteresting to the physical cosmology community.

## 6.10  Resolution

Two research programs in cosmology at the turn of the century settled many of the debates of the 1990s. One extended the observations of Hubble's law, the redshift-magnitude relation between distance and redshift (discussed in Sec. 4.2), to distances so large that the recession speeds are relativistic and the predicted relation between redshift and apparent magnitude depends on the curvature of spacetime.[12] The other mapped the pattern of the distribution of the fossil radiation across the sky. The $\Lambda$CDM theory fits the measurements from both of these two quite different ways to probe the universe, as established by the tests to be described here.[13] The consistent results from these two quite different programs drove general acceptance of the $\Lambda$CDM theory. Let us consider some details.

### 6.10.1  THE REDSHIFT-MAGNITUDE RELATION

Reliable application of the first of these experiments, the measurement of the redshift-magnitude relation, requires identification of

12. The astronomers' apparent magnitude in the redshift-magnitude relation is defined in footnote 7 on page 99. At distances great enough that the redshift is large, the recession velocity derived from the observed redshift by application of the Doppler shift should be considered a nominal value, because there are relativistic effects to be considered. The redshift $z$ of an object is defined by the ratio of the wavelength $\lambda_{obs}$ of a feature in the observed spectrum to the wavelength $\lambda_{lab}$ of the feature measured in the laboratory. The redshift is $z = \lambda_{obs}/\lambda_{lab} - 1$. The subtraction of unity is an historical accident, but convenient. When the redshift $z$ is small it is the fractional shift of the wavelength produced by a Doppler shift at speed $v = cz$, where $c$ is the speed of light.

13. The details of the experimental programs are best left to the book, *Finding the Big Bang* (Peebles, Page, and Partridge 2009), and Chapters 8 and 9 in *Cosmology's Century* (Peebles 2020).

objects at great distances that are luminous enough to be seen, and have close enough to the same intrinsic luminosities that their apparent magnitudes at different redshifts may be compared to predictions. Allan Sandage (1961) wrote a celebrated paper on this subject, "The Ability of the 200-Inch Telescope to Discriminate between Selected World Models." Sandage pointed out that this great telescope at Palomar Observatory in Southern California is capable of detecting the most luminous galaxies far enough away that their apparent velocities of recession approach relativistic speeds. This means measurements of redshifts and apparent magnitudes could distinguish between the steady-state cosmology and the relativistic Einstein-de Sitter model, for example, provided that the galaxies observed at different redshifts could be brought to a common standard of intrinsic luminosity. This is a given in the steady-state cosmology, because galaxies at different redshifts always are statistically the same, but it is a real problem in the big bang theory because distant galaxies are observed when they were young, an effect of the light travel time. Comparing distant galaxies to closer, older, ones requires correction for the effects on their luminosities of the formation and evolution of their stars, which is a complicated task.

Supernovae of a particular kind, exploding white dwarf stars, proved to be suitable for this test. Their peak luminosities differ, but can be brought to a common standard by the observed correlation of luminosity with the rate of change of luminosity. In astronomers' terms, a supernova absolute magnitude, a measure of its luminosity, is corrected to a common value by the observed rate of change of its apparent magnitude. The measurements of the redshifts and corrected apparent magnitudes of supernovae at a range of distances can be compared to the predictions of cosmological models.

A tight measurement of this redshift-magnitude relation at redshifts near unity has been a goal for cosmology since the 1930s. Two groups at last achieved it, seven decades later, by observations of these supernovae. It was not a Merton multiple; the groups were in tight competition. Either group could have established the conclusion, but the consistency of results of this difficult measurement by two groups added to the credibility. The measurements fitted to the relativistic cosmological model indicated that the mean mass density of our universe is about 30% of Einstein-de Sitter, and that a cosmological constant keeps space sections close to the flat case most felt is required by cosmological inflation. This achievement was recognized by the award of the Nobel Prize in Physics to leaders of the two groups: Saul Perlmutter, Adam Guy Riess, and Brian Schmidt. Perlmutter and Schmidt (2003) review the results and introduce the literature on all the work that was done to get there.

The results from this redshift-magnitude measurement are credible, but one must be wary of possible systematic errors hidden in the subtle hazards of the measurements and their analyses. Astronomers have long worried about the comparisons of nominally similar objects at different redshifts, meaning at different stages of evolution in a big bang cosmology. The supernova teams of course gave careful attention to this, and outside experts agree their case is good, but a full proof of control for evolution is not possible. It was important therefore that the same conclusion followed from another quite different kind of measurement, one with its own dangers of systematic error, but errors that surely would be different from what one might imagine in the supernova observations.

### 6.10.2 PATTERNS IN THE DISTRIBUTIONS OF MATTER AND RADIATION

Section 5.4 reviews the realization that we are in a sea of radiation, a fossil from the early hot stages of expansion of the universe. Its thermal spectrum and smooth distribution across the sky is consistent with the idea that the universe in its early stages of expansion was very close to homogeneous, uniform throughout the space we can observe. In the established relativistic cosmology, small initial departures from exact uniformity grew by the attraction of gravity into the clumpy distribution of matter seen around us in galaxies and groups and clusters of galaxies. General relativity gives us a theory of the gravitational rearrangement of the matter distribution as cosmic structure grew, and the resulting gravitational disturbance to the distribution of the fossil radiation. In this theory the predicted statistical patterns in the angular distribution of the fossil radiation, and the spatial distribution of the galaxies, carry information about initial conditions and the growth of structure as the universe evolved.

A challenge for this program is that the formation of the galaxies and their clumpy distribution must have disturbed the sea of fossil radiation in ways that are not easy to analyze. As the radiation propagated through space it encountered plasma, as in galactic winds and explosions, which heated and dragged on the radiation. It encountered departures from a smooth spacetime caused by mass concentrations that may be close to static or may be rapidly changing, as in explosions that also more directly disturbed the radiation. If energetic enough, explosions also would have rearranged the matter, complicating or even obscuring the information we hope to find in the space distribution of the galaxies. In effect, we must add to the list of assumptions of the ΛCDM theory in Section 6.8 the hope that the advanced stages of cosmic structure formation were not too violent.

If we can ignore these complications then the predictions of the ΛCDM theory can be trusted, because they are derived by perturbation theory, which looks quite reliable for this purpose, applied to physical processes that are well understood. And the predictions are compared to measurements that we can be sure are reliable because they are well checked by several groups. But there are the two conditions, that violent events of some sort did not seriously disturb the matter and radiation, and that we can rely on fundamental physical theory. This is asking a lot, which means a persuasive case for the theory requires many tests. Let us consider now the argument that there are many tests.

Figure 6.1 shows how the statistical fluctuations in the angular distribution of the fossil radiation vary with angular scale, and how the statistical fluctuations in the spatial distribution of the galaxies vary with spatial scale. The curved lines are the ΛCDM predictions, the results of the same theory compared to the reductions of these two quite different data sets. (The ΛCDM parameters are slightly different in the two panels, because there are small differences of opinion on the measured values, but they are close enough to equal for our purpose.)

The waves in the theory and measurements in both panels are remnants of oscillations of the baryonic plasma and radiation in the early universe, understood as follows. The radiation in the early

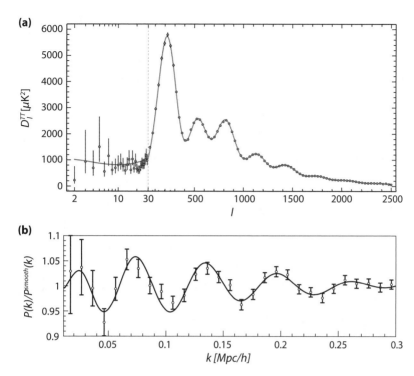

FIGURE 6.1. Signatures of primeval pressure or acoustic waves imprinted on the angular distribution of the fossil radiation, in Panel (a), and on the spatial distribution of the galaxies, at redshifts $0.4 < z < 0.6$, in Panel (b). The solid curves are the predictions. Panel (a) is from the Planck Collaboration (2020), reproduced with permission © ESO. Panel (b) was made by Florian Beutler from data from the Baryon Oscillation Spectroscopic Survey collaboration published by Beutler, Seo, Ross et al. (2017).

universe was hot enough to have kept the baryons ionized until the radiation temperature dropped with the expansion of the universe to about 3000 K. This is redshift $z \approx 1000$. Prior to that, when the baryons were ionized, scattering of the radiation by free electrons, and scattering of the electrons by ions, caused radiation and plasma to move together, acting as a compressible fluid, the radiation providing the pressure. Small departures from exact homogeneity caused this fluid to oscillate in what we can term pressure, or acoustic, or sound waves. This behavior abruptly ended at redshift $z \approx 1000$, when the radiation had cooled enough to allow the plasma to combine to neutral atoms. That freed the baryons from the radiation, leaving both with the traces of the pressure oscillations seen as the patterns of waves in the figure. This was predicted well before it was observed, in the theory developed by Peebles and Yu (1970).

The angular scale in Panel (a) in the figure is $180°/\ell$, where the spherical harmonic index $\ell$ is the horizontal axis in Panel (a). The linear scale in Panel (b) is $\pi/k$, where the wavenumber $k$ is the horizontal axis in Panel (b). The power spectrum in this bottom panel is the square of the Fourier transform of the galaxy space distribution, after taking out the slow variation of the spectrum with wavenumber. Panel (a) is based on the analog of the Fourier transform for data on the sphere. It is the square of the spherical harmonic transform of the fossil radiation temperature. Where the power spectra are relatively large the fluctuations of the radiation temperature or galaxy counts are particularly large when smoothed over angular scale $180°/\ell$ or linear scale $\pi/k$.

The small circles in both panels are the measurements with their uncertainties indicated by the vertical lines. The measurements of the distribution of the fossil radiation on large angular scales, meaning small values of the index $\ell$, are uncertain because at large scales there are not many statistical samples of the pattern across the sky. The measurements are much tighter at larger $\ell$, smaller scales, where there are many samples. These tight measurements closely follow the details of the prediction plotted as the solid curve in Panel (a). This is an impressive result.

The measurements of the statistical pattern in the galaxy distribution are less tight, because there are fewer galaxies than samples of the fossil radiation across the sky, but the consistency of theory and observation for this statistical measure of the galaxy distribution is clearly seen. This consistency also is an impressive result.

It is even more impressive that the same ΛCDM theory and parameters fit the statistical measures obtained by these two quite different means of probing the universe, by the observations and analyses of the distributions of matter and of radiation.

There are other tests, among them the measurements to be discussed in the next Section 6.10.3. But Figure 6.1 makes the central point: the ΛCDM theory passes a demanding test from the consistency of theory and observations of the statistical patterns in the distributions of galaxies and the fossil radiation. Might a physically different theory be consistent with the measurements illustrated in

**Table 6.1.** Checks of Cosmological Parameters

| | matter $\Omega_m$ | baryonic $\Omega_{baryon}$ | helium $Y$ | Hubble's constant, $H_0$ | age of the universe |
|---|---|---|---|---|---|
| $z \lesssim 0.1$ | $\sim 0.3$ | $\sim 0.05$ | $0.245 \pm 0.004$ | $74.0 \pm 1.4$ | $13.4 \pm 0.6$ |
| $z \sim 1$ | $0.26 \pm 0.06$ | — | — | — | — |
| $z \sim 10^3$ | $0.32 \pm 0.01$ | $0.049 \pm 0.001$ | $0.24 \pm 0.03$ | $67.4 \pm 0.5$ | $13.80 \pm 0.02$ |
| $z \sim 10^9$ | — | $0.044 \pm 0.001$ | $0.24 \pm 0.01$ | — | — |

Figure 6.1? It can never be disproved, but the idea certainly seems unlikely.

### 6.10.3 QUANTITATIVE TESTS

Table 6.1 summarizes tests from comparisons of different and independent constraints on the values of the parameters in the $\Lambda$CDM theory. The rows are ordered by the epochs back in time to the occurrences of the physical processes that produced the phenomena that were observed and used to deduce the quantities in the table. The epoch is labeled by the redshift $z$, where the factor by which the universe has expanded since the epoch in question is $1 + z$. A nominal choice in the first row, to represent the present epoch, is $z \lesssim 0.1$, which is the last ten percent of the expansion of the universe to date. At the epoch $z \sim 1$, in the second row of the table, the mean distance between galaxies was half the present value. At redshift $z \sim 10^3$ there likely were no galaxies, so let us think of the mean distance between conserved protons and neutrons. They were on average a thousand times closer together at $z \sim 10^3$ than they are now. The particles at $z \sim 10^9$ were a thousand million times closer than at the present epoch. We pay attention to these enormous extrapolations back in time because they contribute to the demonstration of a consistent story of cosmic evolution.

The evidence is that space sections are close to flat, consistent with the vanishing space curvature expected in the usual interpretation of cosmological inflation, so for simplicity I shall stipulate that the version of the $\Lambda$CDM theory we are testing has flat space sections. Consider it an additional assumption, one that simplifies the discussion without obscuring the point to be argued. The value of Hubble's constant, $H_0$, enters some measures. As another short-cut I use $H_0 = 70$ km s$^{-1}$ Mpc$^{-1}$ when not otherwise available. (In a convenient abbreviation, this is $h = 0.7$.) We can be reasonably sure this is not far off.

The value of the parameter $\Omega_m$ at the head of the second col-umn in the table is the ratio of the present cosmic mean mass density to the mass density in the Einstein-de Sitter model that is expanding at escape speed at the observed rate of expansion of the universe.[14] The straightforward reading of the evidence since the 1980s, from observations of what is happening at the present epoch or close to it, has been that the mass density is about a third of the Einstein-de Sitter value, $\Omega_m \sim 0.3$. This evi-dence was not welcome in the years around 1990, because in general relativity it requires either space curvature, which is con-trary to the usual reading of inflation, or a cosmological constant, which we have seen leads us to deeply puzzling issues. By the turn of the century the tests required acceptance of the cosmological constant with close to the flat space sections adopted to simplify the table. The local observations continue to support the evidence that the present mean mass density is close to the entry in the table.[15]

14. In the general theory of relativity the rate of expansion of the homogeneous uni-verse as a function of the cosmological redshift $z$ is given by the Friedman-Lemaître equation. It is well approximated as

$$\frac{1}{a}\frac{da}{dt} = -\frac{1}{1+z}\frac{dz}{dt} = H_0 \left[ \Omega_m(1+z)^3 + \Omega_r(1+z)^4 + \Omega_\Lambda + \Omega_k(1+z)^2 \right]^{1/2}.$$

The expansion factor $a(t) \propto (1+z)^{-1}$ is the scaling with time of the mean distance between conserved particles. On the left-hand side of this expression are two ways to write the rate of expansion of the universe at redshift $z$. The present rate of expansion, at $z = 0$, is Hub-ble's constant, $H_0$. The dimensionless parameters $\Omega_i$ are the fractional contributions to the square of the present expansion rate by different components: $\Omega_m$ represents the mass den-sity in baryonic plus dark matter, $\Omega_r$ the mass density of the fossil radiation accompanied by the assumed massless neutrinos, $\Omega_\Lambda$ the contribution by the cosmological constant, and $\Omega_k$ shows the effect of the curvature of space sections. The four dimensionless parameters add to unity: $\Omega_m + \Omega_r + \Omega_\Lambda + \Omega_k = 1$.

15. Here are collected references for the table. For $\Omega_m$, at low redshift: Table 3.2 in Peebles (2020), and Abbott, Aguena, Alarcon et al. 2020; at $z \sim 1$: Tonry, Schmidt, Barris et al. (2003), and Knop, Aldering, Amanullah et al. (2003); at $z \sim 10^3$: Planck Collabo-ration (2020, hereinafter PC20). For $\Omega_{baryon}$, at low redshift Schellenberger and Reiprich

The value of $\Omega_m$ in the second row of the table was derived from the program of supernova redshift-magnitude measurements discussed in Section 6.10.1. The form of this relation is sensitive to the value of the matter density at $z \sim 1$, when the universe was about half its present size, and the rate of expansion of the universe was changing from deceleration due to the gravitational attraction of the matter to acceleration due to the repulsion effect of the positive cosmological constant.

The value of $\Omega_m$ in the third row is based on the theory of what happened when the fossil radiation temperature dropped to $T \sim 3 \times 10^3$ K, becoming cool enough to allow the baryonic plasma to combine to neutral atoms, releasing the baryons from the grip of the radiation (discussed in Sec. 6.10.2). This is known as decoupling. The abrupt end of the fluid-like phase for the baryons and radiation produced the statistical patterns in the distributions of galaxies and the fossil radiation shown in Figure 6.1. The detailed shapes of the patterns depend on the mass density in matter, because that helps determine the time of expansion when the baryons and radiation decoupled. Just as the length of an organ pipe determines its pitch, the quasi-wavelengths of oscillations in Figure 6.1 depend on the speed of sound and the time of expansion from high density to decoupling.

Another measure of $\Omega_m$ is from the effect on the fossil radiation as it passes through the evolving gravitational potential of the growing concentrations of mass. This is detected by a correlation between the observed concentrations of galaxies at redshifts $z < 0.8$ and the departures of the radiation temperature from uniformity: $\Omega_m = 0.30 \pm 0.01$ (Hang, Alam, Peacock et al. 2021).

Some of these measures of $\Omega_m$ are not very precise, but all seem reliable. Their consistency within the uncertainties is impressive because they are based on observations of phenomena that were produced by what was happening at quite different epochs in the course of evolution of the universe: now, from the dynamics of

(2017), and Macquart, Prochaska, McQuinn et al. (2020); at $z \sim 10^3$: PC20; at $z \sim 10^9$: Cooke, Pettini, and Steidel, (2018). For the helium mass fraction $Y$, at low redshift: Aver, Berg, Olive et al. (2020); at $z \sim 10^3$: PC20; at $z \sim 10^9$: PC20. For Hubble's constant $H_0$, Di Valentino, Anchordoqui, Akarsu et al. (2020a). For the age of the universe, Di Valentino, Anchordoqui, Akarsu et al. (2020b), and PC20.

relative motions of the galaxies; back in time to $z \sim 1$, when the universe was half its present size, from the redshift-magnitude measurements and the effect of the growing mass concentrations on the radiation; and still further back to $z \sim 10^3$, when baryonic matter was freed from the radiation and the signatures of the pressure oscillations were imprinted on both. Consider also that the measurements of these quite different probes of what happened at well-separated stages in the evolution of the universe were obtained by different methods of observation. The data were reduced by application of the ΛCDM theory. If there were something seriously wrong with this theory, or if the observations were seriously flawed, we would not expect the three numbers to have anything much to do with each other. Might the consistency of the values of $\Omega_m$ be only accidental, a Putnam miracle? It certainly looks unlikely. Instead, this consistency of estimates derived from such different ways to look at the universe makes an excellent empirical case for the relativistic hot big bang ΛCDM theory of the evolution of the universe.

A century ago Peirce pointed to a similar situation for the speed of light, $c$ (reviewed on pages 4 to 9). By my count Peirce listed four ways to measure $c$, by results from astronomical observations and laboratory measurements. The precision of these measurements has improved since then, and we have a better theory; relativity replaced classical mechanics. I expect the three values of $\Omega_m$ in the table will be adjusted as the measurements improve, and I hope they will be adjusted when a better theory than ΛCDM is found. But Peirce's point remains. The consistency of results from different ways to probe the world around us, as in the four measures of the speed of light in Peirce's time and now the three measures of $\Omega_m$, is what one would expect if the theories used to reduce the observations were useful approximations to the way the world around us operates.

The parameter $\Omega_{\text{baryon}}$ in the third column of the table is the contribution to $\Omega_m$ by the mass density of the baryonic matter. In the standard theory the rest is the nonbaryonic cold dark matter introduced to cosmology in 1982. We have two estimates of $\Omega_{\text{baryon}}$ from observations at close to the present epoch. One is based on measurements of the baryon mass fraction in rich clusters of galaxies. These clusters are thought to be large enough to have captured a fair sample of the baryon mass fraction, which is reasonably well supported by numerical simulations. Most of the cluster baryons are in the intracluster plasma, and X-ray spectrum measurements give reasonable measures of the plasma mass. The mass in baryons in stars in a cluster is less certain, but this is a smaller contribution to the total. The mass in baryons plus dark matter is derived from the gravity needed to hold the cluster together. The ratio of the cluster mass in baryons to the total cluster mass is an estimate of the cosmic baryon mass fraction, $\Omega_{baryon}/\Omega_m$. That multiplied by $\Omega_m$ from local dynamical observations gives an estimate of $\Omega_{baryon}$. The second measurement at close to the present epoch is based on the observations of fast radio bursts received from other galaxies. The free electrons along the line of sight through intergalactic space cause the pulses detected at longer wavelengths to arrive later. (The time delay is proportional to the square of the wavelength, and it is a measure of the free electron density integrated along the line of sight from the source to observer: electrons per unit area.) The time delay is not affected by electrons bound to atoms, but the quite small absorption of ultraviolet light by neutral atoms along the lines of sight from distant galaxies shows that intergalactic matter is almost entirely ionized. Time delays as functions of wavelength have been measured for galaxies as distant as redshift $z \sim 0.5$. I lump the results in the row $z \lesssim 0.1$.

These two estimates of $\Omega_{baryon}$ at relatively small redshifts are not very precise, but both are thought to be reliable, and the results are consistent. This near local baryon density is a valuable benchmark to compare to the baryon density inferred by other more indirect methods.

The estimate of $\Omega_{baryon}$ at $z \sim 10^3$ is derived from the statistical measures in Figure 6.1 of the patterns imprinted at baryon-radiation decoupling on the distributions of matter and radiation. The mass density in the baryonic plasma that was coupled to the radiation helps determine the speed of pressure waves, which helps determine the detailed shapes of the curves in the figure.

The measure of $\Omega_{baryon}$ in the bottom row of the table traces the expansion of the universe back by a factor of $z \sim 10^9$. This is when the theory has it that thermonuclear reactions produced the bulk of the present isotopes of hydrogen and helium. The predicted abundance of deuterium, the stable heavier isotope of hydrogen, is sensitive to the baryon mass density: the greater the baryon density at $z \sim 10^9$, the more efficient the fusing of deuterons to produce heavier isotopes, and the lower the residual prestellar deuterium abundance. The best measurements of this prestellar deuterium are from observations of small young galaxies at redshifts $z \sim 2$ to 3. They have low abundances of the heavier chemical elements that are produced in stars, and the presumption is that this means the deuterium abundance has been little affected by what the stars have been doing. The deuterium abundances in these galaxies are measured by the deuterium absorption lines, which are shifted from the hydrogen lines because a deuteron has twice the mass of a proton.

The measurements of $\Omega_{baryon}$ at $z \sim 10^3$ and $z \sim 10^9$ differ by more than the nominal uncertainties because of the Hubble parameter discrepancy to be discussed. The measure at $z \sim 10^3$ uses the value $h = 0.67$ derived from the patterns imposed at decoupling. The measure at $z \sim 10^9$ uses the nominal value $h = 0.7$. But the discrepancy is modest considering the great extrapolation back in time.

Let us recite the key point. We have three measures of the baryon mass density. The first, the density in the universe as it is about now, is based on observations of the baryon mass fraction in clusters of galaxies and on measurements of the time delay of fast radio bursts caused by the free electrons in intergalactic space. These independent ways to look at the baryon density agree. The second is derived from the theory of decoupling of baryons and radiation at redshift $z \sim 10^3$ and the formation of the observed statistical patterns in the distributions of radiation and galaxies. The third is the inference from the observed abundance of deuterium relative to hydrogen interpreted by the theory of light element formation in the much earlier universe, redshift $z \sim 10^9$. These estimates of $\Omega_{baryon}$ are based on different theories of what happened at three well-spaced epochs in the evolution of the universe, observed in three different ways. If the theory were wrong, or the observations misinterpreted, we would not expect the reasonable consistency of results entered in the table. It is Peirce's (1878a) point again. The values of $\Omega_{baryon}$ in the table are sure to change as the measurements improve, and maybe when a better theory is found, but the picture of a relativistic hot big bang with its dark matter and cosmological constant does not look likely to change much.

The fourth column of the table shows estimates of the abundance $Y$ of helium, measured as a fraction of the baryonic mass. The entry in the first row, at $z \lesssim 0.1$, is from observations of helium recombination lines in astrophysical plasmas. The entry in the third row follows from the theory of the effect of the helium on the speed of sound in the plasma-radiation fluid, which affects the patterns in the distributions of baryonic matter and radiation frozen in at decoupling at $z \sim 10^3$. The helium abundance figures in a quite different way in the theory of light isotope formation at $z \sim 10^9$, for the entry in the fourth row.

Let us recite the key point yet again. We see a reasonable degree of consistency of these three estimates of the helium abundance $Y$. They are based on what happened in three different observable ways in the course of evolution of the universe, and observed by quite different methods. This consistency of estimates of $Y$ by itself is a good argument for the $\Lambda$CDM theory. This with the measurements of $\Omega_m$ and $\Omega_{\text{baryon}}$ makes a persuasive case.

The value of Hubble's constant, $H_0$, in the fifth column and first row of the table, is a measure of the present rate of expansion of the universe, from observations of galaxy distances and their redshifts. The value in the third row is inferred from the pattern of the angular distribution of the fossil radiation set at decoupling. Both are considered to be reliable to a few percent. At the time of writing the two differ more than expected from a careful assessment of the measurement uncertainties. Community opinion about this is of three minds.

Some argue that the $\Lambda$CDM theory passes so many other tests that we might expect it to pass this one too; surely the error is in the measurements. But the measurements have been well checked, so many argue that the discrepancy is a real anomaly. Many, including me, hold to both of these opinions along with a third. The universe expanded by a factor of 1,000 between the processes at $z \sim 10^3$ and at $z \sim 0$ that produced phenomena that indicate values of $H_0$ that differ by about five to ten percent. A 10% discrepancy from tracing the expansion of the universe back in time by a factor of 1,000 is impressively good. It is what would be expected if the ΛCDM theory were a good approximation to reality, as indicated by the checks of $\Omega_m$, $\Omega_{baryon}$, $Y$, and $H_o$, but that the approximation stands to be improved.

An improvement of the theory certainly would not be surprising. In particular, the pictures of the dark matter and the cosmological constant look seriously schematic, as if they were only the simplest approximations we could get away with at the present level of the evidence. The overall success of the tests we have considered leads me to anticipate that, while it is easy to imagine the theory will be improved, it is not at all likely to be overthrown. If it were, what would we make of the many positive results of these tests? The likely interpretation is that the ΛCDM theory requires adjustments. I place my bet on a better model for the dark sector, something that behaves in a more interesting way than dark matter and Λ.

The last column in the table shows the age of the universe, computed from the time when it was really hot and dense. The entry in the first row is from the theory of stellar evolution applied to the oldest stars in our galaxy. Stars are observed in other distant galaxies, at redshifts approaching $z \sim 10$, which would be when the universe was young. This suggests the oldest stars in our Milky Way galaxy might be only slightly younger than the time of expansion of the universe from high density. The third row is inferred from the theory of what happened to the fossil radiation at decoupling, $z \sim 10^3$. The situation at the time of writing is that these two quite different ways to estimate or constrain the cosmic expansion time agree. This is yet another probe of the universe that yields results consistent with ΛCDM, to be added to all the other evidence in Table 6.1.

The more direct tests of relativity are important too, because our physical cosmology is based on Einstein's general theory of relativity. But the situation already is clear: ΛCDM passes an impressive variety of tests. Let us go on to consider lessons about the nature of physical science and the assumption of objective physical reality.

# CHAPTER SEVEN

# Lessons from
# a Scientific Advance

The development of cosmology gives us a worked example of the ways of research in physical science. It is the familiar mix of good management and good luck: guidance from the phenomena and what has worked in other situations, intuition and shrewd guesses, and mistakes made and, largely, corrected. The $\Lambda$CDM theory is not complete, but that is true of all our theories. This theory has been checked by the tests of predictions reviewed in Section 6.10. They are abundant enough to make a persuasive case that $\Lambda$CDM is a good and reliable approximation. The case for a physical theory is never proved, of course, but this one is compelling for most of us who are familiar with the subject.

Some theories are tested by laboratory experiments in carefully controlled situations. Physical cosmology is tested by observations that cannot be controlled but can be interpreted with some confidence because the situations look simple. This apparent simplicity is tested, along with the theory, by the reproducibility of observations and their consistency with predictions of the theory.

The emphasis on controlled experiments, or simplicity of physical situations, is required at the present state of the art of physical science, but it is in striking contrast to the complexity of the world around us. Checking whether this has biased our physical theories is a fascinating challenge that will be addressed by progress in learning how to deal with complex physical situations. To be seen is how well this can be done.

## 7.1   Discovery of the ΛCDM Theory Seems Inevitable

Community acceptance of the idea that most of the mass of the universe is nonbaryonic might be counted as a paradigm shift of the sort Kuhn (1970) introduced. Indeed, when astronomers and cosmologists think about a galaxy we are conditioned now to picture a cloud of dark matter that contains traces of baryons: stars, gas, and plasma. But everyone still is very interested in baryons, and the dark matter is better described as a paradigm addition, one of the many steps toward the construction and testing of the ΛCDM world picture.

Was the construction and acceptance of the ΛCDM theory inevitable? Or was it an eventuality, an accident that happened to direct attention to this theory rather than something completely different? An exchange of nuclear weapons with the USSR over the crisis in Cuba in the early 1960s could have destroyed the chance of our ever finding this theory, as well as so much more. But given the state of technology in 1960, and the relative stability of the cultures of science and society that went with it, ΛCDM seems sure to have been discovered. To argue for this I offer the multiple paths that could have taken us to this theory, starting with the idea of a hot early universe.

The two early empirical clues to the hot big bang theory were the sea of thermal radiation and the large abundance of helium. We have a convincing case now that both are remnants from the hot early stages of expansion of the universe. If Bob Dicke had continued research in laboratory quantum optics instead of forming the Gravity Research Group, then others familiar with the astronomical evidence of a curiously large abundance of helium still would have recognized the possible connection to George Gamow's hot big bang theory. Donald Osterbrock and Geoffrey Burbidge saw it in the early 1960s. Fred Hoyle's independent recognition in 1964 was better advertised, and difficult to overlook. If Hoyle's attention had been directed instead to the structure and evolution of stars, another of his interests, the astronomers' growing evidence of large helium abundances in old stars with low abundances of the heavier chemical elements nevertheless was growing difficult to overlook. Fred Hoyle and Martin Schwarzschild understood the issue, an interesting cosmological test, in 1957 (as described on page 113). News of the large helium abundance surely would have reached Yakov Zel'dovich in Moscow and corrected his thinking that there is little helium in old stars. Given this he and his group were well prepared to pursue the idea of a hot big bang. If Gamow in 1948 had been thinking about the genetic code instead of cosmology, someone else could have found the hot big bang theory. I did, independent of Gamow, at Dicke's suggestion. Yuri Smirnov in Moscow certainly was in a condition to have done the same, without the hint he had from Gamow but with the hint of a curiously large helium abundance.

The fossil radiation from the hot big bang was detected in Bell Laboratories' microwave communication experiments, at least as early as 1959. If Arno Penzias and Robert Wilson had not made a determined effort to find the source of the excess radiation five years later, then the search for the fossil radiation by the Princeton group would have detected it within the year. Absent those two groups, the engineering of microwave communication, as in mobile phones, could not have gone on much longer without public recognition of the pesky detection of excess noise. As with the excess helium, there were people capable of seeing that this excess noise might be remnant from the hot big bang. I include members of Zel'dovich's group in Moscow, Dicke's group in Princeton, and Hoyle with colleagues and students in Cambridge. Hoyle knew the puzzling detection of the interstellar cyanogen molecule in its first excited energy level, as if these molecules were excited by the presence of a sea of microwave radiation. McKellar, at the Dominion Astrophysical Observatory in Canada, put the effective temperature at a few degrees Kelvin. When the excess noise in the Bell receivers, a few Kelvins, was announced many years later, astronomers with remarkable memories promptly pointed out the possible connection. One of these astronomers could have been first to alert the community to this evidence of a sea of radiation. Hoyle could have done it; he knew the evidence from interstellar cyanogen, though he had forgotten it in the 1960s.

Given Hoyle's distrust of the big bang picture he might be expected to have suggested that the source of this microwave radiation is a new class of microwave-luminous galaxies. In this alternative or counterfactual history, as in the real one, curiosity-driven astronomers would have made a determined effort to see whether this radiation breaks up into compact sources in galaxies when observed with better angular resolution. This happened in observations of the longer wavelength radio background, but not here. Curiosity-driven research being what it is, the intensity spectrum of the microwave radiation surely would have been measured, and found to be close to thermal. It is a sea of radiation with a simple interpretation, a remnant of a hot big bang that would have left a lot of helium.

This simple interpretation of the helium abundance assumes the general theory of relativity. If in this counterfactual history Einstein had decided to become a musician then the consensus in the fundamental physics community is that general relativity would have been discovered by the more pedestrian methods of classical field theory with the condition of local conservation of energy and momentum (Feynman, Morinigo, and Wagner 1995).

I offer this list to show that there were multiple paths to the hot big bang picture. It is important also that it is difficult to see the possibility of interesting detours that could have led us to some picture other than a hot early state of a relativistic expanding universe. Might my difficulty in spotting interesting detours be a conditioned limit on my imagination? It is quite conceivable as I am now. But there are and have been hungry young people like I once was who would have been delighted to point out these clues overlooked by the establishment.

There also were multiple paths to subluminal matter. Fritz Zwicky saw the problem with the mass of the Coma Cluster in 1933. In 1939 Horace Babcock came across unexpectedly large motions of plasma in the outer parts of the galaxy M 31, a hint to the subluminal mass that helps hold the galaxy together. The systematic exploration of the properties of galaxies led to Gérard de Vaucouleurs' report of extra mass relative to starlight toward the edges of M 31. It was based on measurements in the Netherlands of the motions of atomic hydrogen and the inferred mass distribution in this galaxy, and de Vaucouleurs' measurements of the distribution of starlight. This was in 1958, two decades after Babcock. A decade after that, in 1970, Ken Freeman published evidence that much of the mass of another spiral galaxy had to be outside the luminous region, based on his own method of analysis of 21-cm observations. In the same year, Vera Rubin and Kent Ford (1970a,b) reported their greatly improved measurements of the motion of plasma in the disk of M 31.

Rubin and Ford had the advantage of Baade's identifications of emission line regions, but Rubin (2011) recalls that they had begun their project before learning of Baade's finding list. Rubin and Ford had a new generation of more sensitive detectors. Red-sensitive photographic plates alone could have gotten them their result without the finding list, far more slowly, but astronomers can be persistent. Their precision rotation curve grew out of the curiosity-driven research that is the wont of competent observers. They had no particular interest in physical cosmology, but they gave cosmologists valuable evidence: subluminal matter. The unexpectedly large circular velocities in the outskirts of spiral galaxies were there, sure to be observed if observational astronomers had the support to allow them to follow their desire to observe, and hit on the indication of subluminal matter.

A detour not broadly explored was the consideration of a modification of the inverse square law of gravity. If the law of gravitational acceleration at suitably large separations were inversely proportional to the first power of the separation, rather than the usual square of the separation, it could eliminate the need for the postulate of subluminal matter. This certainly was worth exploring, because the application of the inverse square law to the structures of galaxies is a long extrapolation from its well-tested applications to the dynamics of the solar system (with tiny relativistic corrections). Milgrom (1983) created the most notable idea, his Modified Newtonian Dynamics, MOND. The idea still is discussed, and the search for a cosmology based on MOND, or some other way to avoid the postulate of something that acts like dark matter, is good science. But it looks exceedingly unlikely that a theory without something equivalent to dark matter could pass the multiple tests reviewed in Section 6.10. Consider in particular the oscillations of the power spectrum of the spatial distribution of the galaxies in Figure 6.1. The measured amplitude is small. In the theory this is because the dominant mass in dark matter does not take part in the acoustic oscillations at $z > 1000$. Without something that acts like dark matter, how would this small amplitude be understood? More broadly put, if a theory without dark matter passed the constraints in Table 6.1 as well as $\Lambda$CDM it would indicate a quite implausible number of Putnam's miracles. But as I keep remarking, this is a judgement, not a proof.

The idea that the subluminal matter is not baryonic grew out of the two sets of Merton multiples discussed in Section 6.6: nonzero neutrino rest masses and a fourth kind of neutrino. The multiples tell us that the idea was reasonably sure to be discovered; it only required that someone knowledgeable about the search for completion of the standard model of particle physics recognize the interesting constraint on neutrino rest masses from reasonable bounds on the cosmic mean mass density. We have seen that among those who recognized this at least some knew another interesting piece of evidence: subluminal matter in clusters of galaxies. If they had better understood the astronomers' evidence of subluminal matter around galaxies it would have helped, but it wasn't needed.

The idea of nonbaryonic matter from particle physics was an offer cosmologists could not refuse. The path taken from subluminal matter to the $\Lambda$CDM theory went through Peebles (1982). It was a Merton singleton, but consider that Gunn, Lee, Lerche et al. (1978), and Steigman, Sarazin, Quintana, and Faulkner (1978), already had been thinking about the merits of what became known as CDM as an aid to understanding the properties of galaxies. They knew the big bang cosmology and its phenomenology; members of these groups had written influential papers on the subject. They were capable of putting together CDM and cosmology. Turner, Wilczek, and Zee (1983) independently explored the idea of cosmology with CDM. I mentioned earlier that they knew the particle physics but their less tight grasp of the phenomenology led them to conclude that the CDM idea is not promising. They were capable of learning, of course.

Meanwhile, in the Soviet Union, Zel'dovich and his group were investigating a cosmological model with another kind of nonbaryonic matter, HDM, one of the known kinds of neutrinos with an assumed rest mass of a few tens of an electron Volt. This candidate for non-baryonic dark matter has the serious problem with galaxy formation discussed on page 158. Peebles (1983) and Melott, Einasto, Saar et al. (1983) saw it, independently, a Merton doubleton. Melott et al. recommended replacing HDM with the stable supersymmetric part-ner remnant from an early epoch of supersymmetry. It could serve for CDM as well as a massive fourth neutrino. Melott et al. were in contact with Zel'dovich and his group. The people in the Melott et al. and Zel'dovich et al. groups were quite capable of working out the advantage of changing from the problematic HDM to the promising CDM theory, if I had not done it first. I am not being modest; this is my attempt to see how the story might have played out. It reveals the considerable variety of ways to the discovery of the concept of nonbaryonic matter in the hot big bang cosmology, and scant hints to the possibility of a different conclusion.

In the years around 1990 the idea that the astronomers' sub-luminal matter is the particle physicists' nonbaryonic matter had become the commonly accepted picture in the cosmology commu-nity; it was involved in most of the lively debates about how best to bring ideas and evidence together. There were two good empir-ical reasons to consider this hypothetical notion of nonbaryonic dark matter. First, the theory of element formation in the early stages of expansion of the universe calls for a baryonic mass den-sity on the low side of what dynamical analyses of galaxy motions indicated. Nonbaryonic matter could add to the mass without disturbing the successful theory of formation of the isotopes of hydrogen and helium. Second, the assumption that the mass den-sity in dark matter is larger than the baryonic component allows a simple explanation of why the fossil radiation is so much more smoothly distributed than the matter. This did not add up to a com-pelling empirical case, of course. Nonbayonic matter in 1990 was a social construction with a significant empirical side.

Until the late 1990s we had at least one viable alternative to non-baryonic matter, the PIB model. It was convincingly falsified by its incorrect prediction of the anisotropy of the fossil radiation. But why was the community not more energetically searching for other alternatives to the speculative idea of nonbaryonic matter? I have mentioned (on page 165) the consideration that it is not very difficult to analyze the formation and evolution of galaxy-size clouds of pure CDM, in analytic approximations and numerical N-body simulations. It was something to do, and the results looked promising Add to this social pressure from our herd instinct. Almost everyone who was thinking about physical cosmology in the 1990s was thinking about nonbaryonic matter, so almost everyone continued to think about it.

The community in the 1990s was paying close attention to measurements of the angular distribution of the fossil microwave radiation, because it constrains ideas about how cosmic structure— galaxies and their clustering—could and could not have formed. Well before that, when the presence of this radiation was first recognized, the immediately obvious thing to do was to check whether the radiation is from galaxies. The failure of improved angular resolution to reveal that the sea of radiation breaks up into individual sources, and the demonstration that the spectrum is close to thermal, made the local source idea uninteresting. In the 1990s the search for departures from an exactly smooth sea was driven in part by the idea that the radiation could not have entirely escaped disturbance by the formation of the clumpy matter distribution. Peebles, Page, and Partridge (2009) describe (in Chapter 5 in *Finding the Big Bang*) the international effort devoted to research aimed at checking for the disturbance to the radiation predicted by variants of the CDM model (reviewed in Sec. 6.9). My impression is that an even greater motivation for many working on these anisotropy measurements was the fascinating challenge of designing an experiment capable of detecting a phenomenon that looked likely to be physically interesting, seemed sure to be present, and seemed likely to be just possibly detectable. It culminated in big science, the WMAP and PLANCK satellite missions among others.

The collection of data for Panel (a) in Figure 6.1 was theory-driven, but I think that in an alternative history without the ΛCDM theory the program of measurement and statistical analysis of the angular distribution of the sea of radiation would have been hard to resist. I developed this program of statistical measures of the galaxy space distribution and motion with no motivation apart from the thought that something interesting might turn up. Absent the theory of how the radiation might have been disturbed, the measurements leading up to Figure 6.1 surely would have been made if observers were allowed to indulge in their love of making ever better measurements. The waves in the two-point function based on these measurements are there, to be seen in a commonly used statistical measure. Interpretation of the waves seen for the galaxies in space, as well as the radiation on the sphere, calls to mind pressure oscillations. All this would have been pretty obvious even if not anticipated by my theory. And recall another point: if there were no dark matter these waves would be expected to have had much greater amplitude than observed. The opportunity for a bright young scientist would have been there: solve the problem by postulating dark matter.

The gathering and analyses of data on the relative positions and motions of the galaxies occupied a considerable fraction of the cosmology community in the years around 1990. I refer the reader to the illustration of this activity by the considerable length of the list of measurements in Table 3.2 in *Cosmology's Century* (Peebles 2020). The goal of this research was to improve statistical measures of the cosmic mean mass density, and later to test consistency with the predicted pattern in the galaxy distribution remnant from decoupling. But again I expect the personal imperative for many was to make the best possible measurements of interesting phenomena. This research was small science when I was leading it; none of our papers listed more than four authors. There are some 30 authors in the papers reporting the observations that yielded the data points in Panel (b) in Figure 6.1. These observations were at least in part theory-driven but I think that, absent the theory, curiosity-driven research would in time have led to catalogs of galaxy angular positions and redshifts, and the application of a standard statistic would have revealed those suggestive waves in Figure 6.1.

The winnowing of the considerable variety of ideas that were discussed in the 1970s through the 1990s, during the early stages of construction of physical cosmology, evolved to increasing attention to the few that looked promising, the usual course of events. A signature of this maturing science was the investigation of well-defined issues probed by research programs that the community agreed are worthy of considerable expenditure of resources. Given the willingness of society to support what had become expensive research programs that do not seem at all likely to be monetizable, the remarkable advances in the construction of physical cosmology seem inevitable. Society paid for these programs, but it is difficult to see how society otherwise influenced the results.

We can dismiss the idea that we have misread the evidence for the construction and establishment of the $\Lambda$CDM theory: consider the number and variety of artificial agreements of theory and observation that would require. We might ask instead whether, among the many paths that could have taken us to the $\Lambda$CDM cosmology, there are other paths not taken that would have led us to a different empirically satisfactory cosmology. I have explained in some detail why this looks exceedingly unlikely. I must add that, as a practical matter, the community is not going to turn to the search for another basin of attraction in the phenomenology unless the set of basic assumptions of the $\Lambda$CDM theory is falsified. There is no generally acknowledged hint that this might happen, but time will tell.

There is a tradition of declarations that mathematical theorems, and successful theories such as general relativity and $\Lambda$CDM, were there, waiting to be discovered. My literal mind makes me uneasy about that, because I don't know where the "there" might be. I am comfortable with the assertion that the fossil helium and radiation were there, and the waves in Figure 6.1 were there in a commonly used statical measure. These were instructive phenomena waiting to be discovered. Since curiosity can be a powerful motivation for research, and society is willing to allow some of us to indulge in it, it is reasonably clear from the arguments I have given that we would see how to build on these clues. It sometimes happened multiple times. It brought us to the impressive cross-checks of measurements consistent with theoretical predictions illustrated in Figure 6.1 and Table 6.1.

I began this section with the question: was the construction of the $\Lambda$CDM theory inevitable? The evidence reviewed here makes the case that it would happen about as close to compelling as it can get. And let us note that this is what we expect if our present physical cosmology is a good approximation to objective reality, not a social construction.

## 7.2 Constructions and the Science Wars

The idea of social and empirical constructions brings us to the science wars, the frank and thorough exchanges of views of some sociologists and philosophers on one side, and physical scientists on the other, on what each saw as reasons to object to what they felt they saw the other side doing. Ian Hacking (1999), who "tried to give a fair rhetorical shake to both" sides, gives us an example of the thinking.

> The constructionist maintains a contingency thesis. In the case of physics, (a) physics (theoretical, experimental, material) could have developed in, for example, a nonquarky way, and, by the detailed standards that would have evolved with this alternative physics, could have been as successful as recent physics has been by its detailed standards. Moreover, (b) there is no sense in which this imagined alternative physics would be equivalent to present physics. The physicist denies that. Physicists are inclined to say, put up or shut up. Show us an alternative development. They ignore or reject Pickering's discussion of the continued viability of the old physics.

Hacking is referring to Andrew Pickering's book, *Constructing Quarks* (1984). Allan Franklin (2005, page 210) remarks that

> During much of the two decades covered in Pickering's book, elementary-particle theoreticians were casting about in many directions, and experimenters were correspondingly unsure which experiments to undertake and which parameters to measure.

Social pressures do affect this normal and healthy phase of research. That includes serious considerations of the social constructions that are more charitably termed hypotheses to be tested. I have described the quite schematic empirical situation of general relativity in the 1960s (in Sec. 3.2). That turned out well, as did the situation in particle physics that has grown from the construction phase to a well-defined standard model, a theory. This particle theory has a lot of free parameters, but it produces far more predictions that are definite and pass well-checked tests, making a persuasive case that it too is a useful and reliable approximation to reality.

Might the available phenomenology, maybe gathered in some alternative way, offer another basin of attraction for particle physics or cosmology? Addressing the challenge, "put up or shut up," is not likely to happen, of course. Sociologists and historians are not equipped to do it. Physicists are not about to try unless they find something seriously wrong with their standard models, which looks like an exceedingly dubious proposition. The standard models of relativity and quantum physics are not complete, and they surely will be improved, but they do not seem at all likely to be replaced. This is not a theorem, of course, it is an argument that looks like a very good bet.

Pickering's line of thinking about particle physics, but applied to physical cosmology, is easier to assess because the theory and observations are simpler. We have seen that the case for $\Lambda$CDM is easy to lay out, and that it seems perfectly clear that there were multiple possible paths to a single basin of attraction, this relativistic hot big bang theory.

At the heart of the science wars surely is the difficulty of communication between disciplines with different cultures and traditions. I offer the example of the last sentence in the preface to the second edition of the book, *Laboratory Life*, which reports Bruno Latour's observations while embedded in the Salk Institute for Biological Studies. The philosopher Latour and sociologist Woolgar (1986) wrote that

> Readers tempted to conclude that the main body of the text replicates that of the original are advised to consult Borges (1981).

I imagine this sentence has useful connotations in some cultures, and I suppose Latour and Woolgar have in mind the well respected Argentine author Jorge Francisco Isidoro Luis Borges, but I am not prepared to sort through his impressive literature to find the message. The culture of natural science would have dictated instead a brief declarative statement. The mismatch makes fertile grounds for misunderstandings.

## 7.3 Multiples in the Discovery of ΛCDM

Parts of the discussion in Section 7.1 on the argument for the inevitability of discovery of the hot big bang cosmology are repeated here—they are worth repeating—to illustrate the phenomenon of multiples in scientific discovery (Merton 1961).

In 1948 Gamow published his new physical cosmology, the hot big bang. In the same year, Bondi, Gold, and Hoyle introduced their steady-state cosmology. I have encountered no evidence that each side was aware of the other. These look like two quite independent introductions of influential world pictures.

Not long after that Dicke, in the late 1950s, and Zel'dovich, in the early 1960s, assembled research groups in gravity physics, relativity, and cosmology. I recall Dicke's comment that he was vaguely aware of Hoyle and Gamow, but not very interested. Zel'dovich was interested in Gamow's hot big bang, but thought it must be wrong. I know of no evidence that Dicke and Zel'dovich were aware of each other until later in the 1960s. The decisions by Zel'dovich and Dicke to move to research in this direction were independent, apart from the conditions of society that encouraged it. Their two groups largely independently made major parallel contributions to the development of physical cosmology.

Physical scientists are familiar with the experience of such apparently independent but surely related developments, though they seldom give it much thought. Sociologists recognize it as a phenomenon that I have taken the liberty of naming Merton multiples. I offer four categories drawn from examples from the history of the growth of physical cosmology.

*First Category.* Some multiples have arguably clear origins. The renaissance of cosmology after Word War II was enabled by, I would say driven by, the release of intellectual energy, new technology, and support for pure research. In the US, industry and the military were impressed by physicists' contributions to war research, and were quite willing to fund curiosity-driven research they could readily afford to see what else scientists might come up with.[1] The pre-war investigations of gravity physics and cosmology offered considerable room for improvement by the application of new technology and fresh thinking about physical processes in an expanding universe. Surely Dicke, Gamow, Hoyle, and Zel'dovich needed no more prodding than this to seize the opportunity. It was a Merton quadruple of the first category.

For a second example consider Lemaître's (1927) demonstration that the relativistic solution to an expanding spatially homogeneous universe predicts that the distances and redshifts of the galaxies satisfy Hubble's law: the recession velocity of a galaxy is proportional to its distance. Two years later Hubble announced evidence of Lemaître's relation. What Hubble knew, and when he knew it, is debated, but we can be sure that he was little interested in the general theory of relativity, and very interested in the observational exploration of what he termed the realm of the nebulae. Since the mid-1910s those who were interested in such things had been wondering why the spectra of galaxies tend to be shifted to the red. In the 1920s the increasing fund of evidence on galaxy redshifts and distances, as well as their angular distributions, was making the question pressing. The time was right for the essentially independent and simultaneous presentation of the prediction and the evidence.

*Second Category.* Some Merton multiples call for more subtle interpretations. In 1933, Fritz Zwicky pointed out the problem of missing mass in the Coma Cluster of galaxies. The following year, Enrico Fermi (1934) published his model for the weak interaction

---

1. The curiosity-driven research by Dicke and his Gravity Research Group was supported in part by the United States Army Signal Corps, the Office of Naval Research of the United States Navy, and the United States Atomic Energy Commission. The influential Chapel Hill Conference on *The Role of Gravitation in Physics*, in 1957, was supported by the Wright Air Development Center, Air Research and Development Command, U.S. Air Force. When Gamow and Hoyle met to discuss the idea of a sea of thermal radiation, in about 1956 (as recalled on page 121), Gamow was a guest of the General Dynamics company, with no duties except to stay nearby, in La Jolla California, where he might drop hints to new ideas from pure research. This situation is considered at greater length in Peebles (2017) Section 7.5.

of electrons and nucleons with a newly proposed particle, the neutrino. In the early 1970s it was proposed that one of Fermi's neutrinos, with a rest mass of a few tens of an electron Volt, could be Zwicky's missing mass. The idea came from Hungary and the US, essentially at the same time, independently. It was a contributing factor to the proposal of a fourth kind of neutrino with a much larger rest mass, about $3 \times 10^9$ eV. This idea was presented in five apparently independent publications in 1977. The five papers manifested at best vague interest in what had grown to be the substantial though poorly advertised astronomical evidence of the subluminal matter around galaxies that became known as dark matter. But the idea in these five papers became the prototype for the nonbaryonic dark matter that I introduced to cosmology in 1982 and became central to the now well-tested ΛCDM theory.

Maybe the two publications in the 1930s only coincidently appeared within the space of a year, though it is curious to see the repetition with the Merton doubleton in the early 1970s and the quintuple in the late 1970s. We communicate in many ways: publications, lectures at conferences and centers for research, mail in the past and now email and web sites, and casual remarks and body language. All convey information that may be passed on through many people before reaching someone who happens to be prepared for whatever version of the thinking came through. Notions of zeitgeist and "thoughts in the air" are vague, but the evidence from these examples makes the phenomenon real.

*Third Category.* The many possible paths to the ΛCDM theory from the post-World War II state of science and society are reviewed in Section 7.1. The actual path taken to this major scientific discovery was an eventuality; the discovery was close to inevitable. The reader can find many actual and potential Merton multiples in the course of this discovery.

*Fourth Category.* Since Merton Multiples are coincidental recognitions, I place coincidences involving recognition of phenomena in a category of multiples. We distrust coincidences because they may be purely accidental; they must occur on occasion if many things happened. We pay attention to coincidences involving phenomena because they on occasion prove to be hints to something physically significant. Recall from Chapter 5 and Section 7.1 that, in 1960, some knew that the abundance of helium in planetary nebulae is larger than one might expect, others knew that Bell Laboratories microwave receivers detected more energy than the engineers expected, and still others remembered the unexpected detection of interstellar cyanogen in its first excited level. A half decade later it was recognized that all three seemed likely to be evidence of the same thing, a hot big bang. Are there more such coincidences of phenomena waiting to be recognized in the present state of physical cosmology? It is easier to see them in retrospect.

I place in a subcategory of this fourth kind the coincidences of advances in experimental or observational research programs with a well-defined goal. Two examples played major roles in the development of cosmology. Both took many years of work, yet in both cases the work by two separate research groups reached fruition at essentially the same time. The first was the demonstration that the sea of microwave radiation has a thermal spectrum. This was important because it showed that the radiation very likely is a fossil from the early universe that arrived to us in close to its original condition, apart from cooling by expansion. Two groups in 1990 gave us this evidence. They used the same detector technology, one instrument carried by a rocket, the other by a satellite. The satellite group had been working on making this measurement for 15 years; the rocket group had been working on it longer than that. To avoid possible confusion I emphasize that this was not a case of one group surrendering to the weight of the other's evidence. Both had a clear independent measurement that alone would have made the case. The two measurements could have been accomplished at quite different times, yet they came to the same conclusion in the space of a few months after a decade and a half or more of development. We cannot attribute this coincidence to competition: both groups were working independently as fast as they could.[2]

A second example is the demonstration of persuasive evidence for the presence of a cosmological constant in a cosmologically flat universe. It came from tests based on two different

2. A like example is the completion of the redshift-magnitude test discussed in Section 6.10.1. But this is more complicated because the two groups were in close contact.

kinds of measurements: the redshifts and apparent magnitudes of supernovae, and the angular distribution of the fossil radiation. The phenomena were probed by different methods of observation and analysis that had to be designed to meet quite different challenges. The near simultaneous announcement of results from the two programs, at the turn of the century, abruptly settled the lively debates in the 1990s on the relative merits of the considerable variety of ideas reviewed in Section 6.10. These two quite different research programs required many years of development, and the results could have been reached years apart, but that is not what happened.

It seems unlikely that the near simultaneous appearances of persuasive results in both of these examples were purely accidental, but what might we make of it? A hint might be drawn from the history of discovery and establishment of the $\Lambda$CDM theory. It argues for the manifestation of a kind of collective organization in the communities of physical science that differs from the familiar formal and informal but explicitly declared research collaborations. I have in mind common interests in theories and observations of phenomena that seem relevant to current thinking in a scientific community, shared explicitly or implicitly. This common interest is a social phenomenon: we tend to agree on interesting directions of research without directly consulting each other, maybe for reasons along the lines suggested by the common ocurrence of Merton multiples. On the observational side, more than one group may be paying attention to the same theorists as well as what other observers are doing. I venture to add that all this is what might be expected if the phenomena were presenting useful clues to approximations to objective reality.

## 7.4 Questions

Here is a collection of questions that arise from considerations of advances in the natural sciences in general and in cosmology in particular. To begin, what set people on the paths to the physical sciences that led to the abundance of evidence about the nature of the world around us that is so well described by relativity and quantum physics? In Peirce's time Darwin's arguments for the origin of the species were new and exciting. Peirce suggested that our species might have been conditioned by experience and adaption to look for something like Newtonian mechanics. It seems reasonable: levers function in a reliable, logical, way, a fact on the ground. The possibility of shearing flakes off rocks to make arrow heads was another, as was the flight of an arrow or spear that can be quite reproducible, given the skill. Curiosity about the phenomena of static electricity and natural magnets led to experiments and eventually tightly reproducible facts about electromagnetism. Maxwell's intuition led him to put aside the search for a mechanical model of the luminiferous ether and turn to the field equations. Their consistency with tests in the laboratory and the vast range of practical applications is another fact on the ground. I expect these facts encouraged the idea that the world around us operates by rules we can hope to discover, and I count these rules as approximations to the nature of objective reality.

One might wonder about the consistency of two aspects of the ΛCDM theory. On one hand is the claim that the theory has been persuasively established by passing the impressive array of demanding tests sampled in Section 6.10. But on the other hand the theory is not complete; we cannot say with any confidence what the dark matter is, or what the universe was doing before it was expanding. The two positions are not contradictory. The evidence is that this theory is a useful approximation, good enough for tight tests of predictions, but it is an approximation, with room for improvement. The same is true of all our standard and accepted physical theories; all are well tested and useful, but all are approximations.

What are the places for nonbaryonic dark matter and the cosmological constant in what will have to be an improved fundamental theoretical physics? What would be the conclusion if the theoretical places seemed artificial at best? What would be the conclusion if ongoing searches failed to detect the dark matter, apart from its gravitational effect, and indicated no signature of the nature of the cosmological constant, not even the hint that the effective value of $\Lambda$ is evolving? In this case these two components of the dark sector of ΛCDM might be considered to remain hypothetical, but that is only a name for these two concepts that already pass demanding tests. If no signatures of dark matter and $\Lambda$ are found apart from the gravitational effects it will not weaken the case for the $\Lambda$CDM theory; the evidence we have now is compelling. We have not been issued a guarantee that we can arrive at a better understanding of these components, or of any other aspect of natural science. We are continuing to make progress, though, and there is no reason to expect that this will stop any time soon.

Our existence requires baryonic matter, and baryonic matter brings with it the fascinating mix of fields that account for the existence of neutrons and protons and the ways in which they behave. This physics is advertised to be simple, but in an interesting way that leads to the vast complexity of our world. The dark sector of the $\Lambda$CDM theory is really simple. One component is the cosmological constant, with what appears to be an exceedingly odd numerical value. The other component is a gas of particles that interact with gravity, but weakly if at all with themselves and the baryonic matter and radiation in the sector we so directly experience. Does the great difference between the natures of the physics in the two sectors of $\Lambda$CDM seem reasonable? Maybe we can get away with a simple picture of the dark sector because we know so little about it?

If the dark sector is more interesting than $\Lambda$CDM, different enough to matter, it will be discovered. Discovery may be aided by dark matter detection, in the laboratory or observations of astronomical events; it may grow out of hints from empirical anomalies in continuing tests of the $\Lambda$CDM theory; or it may arise out of a good idea.

It is natural to wonder about the possible significance of apparently curious numerical coincidences. In the $\Lambda$CDM theory about a sixth of the mass of matter is baryonic. Why are there are roughly similar mass densities in the two components, baryonic and nonbaryonic? Of what use is the latter? If the nonbaryonic matter were not present, but all other conditions were the same as in our universe, galaxies would have been different but would have existed, and their stars would have been baryonic, like ours, with baryonic planets for homes for beings like us. We could live with that. Imagine the other extreme, a hypothetical universe that is the same as ours but for one difference, that the mass fraction in baryons is greater than zero but much smaller than in our universe. The radiative dissipation of energy that causes baryons to settle to form stars would be far slower in this universe, but there would be the familiar concentrations of dark matter, and the baryons in them would in time radiatively dissipate energy, contract, and collapse to stars.[3] The stars present in this universe at any time would be far fewer than we observe, but that should not be a problem for the chance of our existence, because our observable universe surely contains far more planets around sufficiently long-lived stars than we need even if the occurrence of life like ours is exceedingly rare. So why the coincidence of similar masses in baryonic and dark matter? Maybe dark matter has an essential place with baryons in some deeper fundamental physics?

Another numerical curiosity is that in the $\Lambda$CDM theory we flourish at a special epoch in the course of evolution of the universe. Until recently, relative to the cosmic scale of time, the rate of expansion of the universe was being slowed by the gravitational attraction of the matter. That changed at cosmological redshift $z \sim 0.7$, when the mass density had decreased enough to have allowed the cosmological constant to start increasing the rate of expansion of the universe. This turning point happened about 6 Gyr ago. The solar system formed 4.6 Gyr ago, not long after the transition to accelerated expansion. A fluke? Or are we just possibly missing something?

Spacetime curvature varies from place to place, largely associated with the irregular mass distribution. These curvature fluctuations are quite small almost everywhere. One indication is the characteristic velocities of stars within galaxies and galaxies within clusters of galaxies, $v \sim 100$ to $1000$ km s$^{-1}$. This corresponds to spacetime curvature fluctuations on the order of $(v/c)^2 \sim 10^{-7}$ to $10^{-5}$. Another is the excellent fit of the theory to the well-measured pattern of the fossil radiation distribution across the sky. The radiation propagating to us is perturbed by the fluctuations of the spacetime curvature it passes through, but the disturbance is tiny,

---

3. Recombination at decoupling would have been less complete than in our universe, leaving a larger residual ionized fraction. Depending of the kinetic temperature of a dark matter halo, collisions of free electrons and protons, or molecular hydrogen formed with the help of the free electrons, would emit radiation that would cause the baryon component to contract, eventually to stars.

detected only by impressively precise measurements. Spacetime is seriously curved around neutron stars, and even more disturbed at black holes, but these are rare exceptions. Our existence requires a galaxy to contain and recycle debris from evolving stars to produce the chemical elements we are made of, but surely not such a smooth, quiet spacetime. Will a deeper physical theory require our tranquility, or is this just another fluke?

What are we to make of the quantum zero-point vacuum energy density? In the standard quantum physics of matter this energy is real and a source of gravity, both of which are experimentally well checked. The same quantum theory applied to the electromagnetic field produces new predictions that pass demanding tests. It seems to be entirely reasonable to expect that the zero-point energy of the electromagnetic field also is real and a source of gravity. If this quantum energy density is independent of the motion of the observer then it acts like the cosmological constant. This situation would be attractive except that the value of this contribution to the cosmological constant would be expected to be some hundred orders of magnitude larger than the total that is observed. Our existence means that something tamed the value of this constant, but without quite eliminating it. Is the only way to account for this the anthropic interpretation, that we live in one of the rare universes in a multiverse that has a cosmological constant with absolute value small enough that we can live with it? If so, why do we find ourselves in a universe with so many galaxies of stars with planets that would be suitable homes for us? It seems excessive for satisfaction of the anthropic principle. Shall we apply the anthropic principle only where needed?

There is good reason to be careful about the big claims of the ΛCDM theory, and to bear in mind the challenges to the tests of this theory. To begin with, we cannot apply the fundamental check of reproducibility to the universe; we have only the one to observe at only one epoch in its evolution. What is more, we observe a tiny fraction of the universe we suppose is out there. We have a probe of spacetime along the thin cone of paths of the radiation and energetic cosmic rays that have reached us from distant parts during the exceedingly short span of time people have been paying attention to such things. We also have the bits of information that have come to us from our spatial neighborhood along timelike paths through spacetime. An example is the light isotopes that formed relatively close to where we are now in space but long ago in time, in the very early universe. Does the information we receive through these exceedingly narrow windows offer a fair sample of the universe? We must rely on the argument of plausibility from the tests of predictability, the arguments reviewed in Section 6.10 for the established theory of the large-scale nature of the universe.

The physical science community has been conditioned by centuries of experience to trust the significance of tests of consistency of theory and practice. But recall that electromagnetism introduced us to relativity physics, which taught us that we must learn to live with revised notions of relative times and positions. Electromagnetism with the theory of heat—thermodynamics and statistical mechanics—forced us to live with the probabilistic nature of predictions in quantum physics. These adjustments of our picture of how reality operates are not considered to violate the working assumptions placed in the box on page 43; the adjustments are interpreted to mean that we are learning how reality operates. But remember Popper's (1935) point, that a theory that can be adjusted to fit whatever is discovered is of no interest, except maybe as a cautionary example of a just so story. Might physical cosmology be dismissed as a just so story? The argument against this degree of adaptability is the immense reach of confirmed predictions displayed in Section 6.10. The case looks compelling, but present research projects designed to improve the cosmological tests, in progress and planned, will be valuable additions to the evidence. I do not anticipate challenges to the basic ideas of the ΛCDM theory, but empirical tests have far more weight than any of our thoughts.

We have seen that establishing the empirical case for physical cosmology has been easy, relatively speaking, because observations and theory are reduced to well-controlled measurements that check reliably derived predictions from the general theory of relativity with the ΛCDM cosmology. Chemistry experiments can be replicated, by skilled chemists, which is an encouraging fact on the ground. But the complexity of chemistry makes the reduction to tests from fundamental principles far more difficult. Might progress in establishing the basis for the concepts of chemistry lead to consistency with present standard physics or, even better, to an improved fundamental theory? It is a fascinating prospect to be checked as methods of *ab initio* chemistry improve.

Bruno Latour and Karin Knorr-Cetina observed research on the properties of products of living matter, which is even more complicated. Will it be demonstrated that these phenomena rest on a fundamental theory that is consistent with or maybe generalized from what we understand now? Exploring this looks really difficult, but I have been surprised so many times at what can be done in science that I would advise you not bet against it, even though the bet does not look at all likely to be settled any time soon.

A century ago people marveled at the ability of our minds to "fathom the properties of real things" (Einstein 1922b), and that it "seems incontestable, therefore, that the mind of man is strongly adapted to the comprehension of the world" (Peirce 1878a). The progress of science since then adds to the examples of this deeply remarkable phenomenon. Maybe understanding it requires a more sound appreciation of the operation of the human brain, I suppose along with the rest of the nervous system, and maybe all the other systems it connects to: the world around us. Roger Penrose (1989, Chapter 9) argues that we have a long way to go to understand the physical basis for all this.

> Is our picture of a world governed by the rules of classical and quantum theory, as these rules are presently understood, really adequate for the description of brains and minds? There is certainly a puzzle for any 'ordinary' quantum description of our brains, since the action of 'observation' is taken to be an essential ingredient of the valid interpretation of conventional quantum theory. Is the brain to be regarded as 'observing itself' whenever a thought or perception emerges into conscious awareness?

Later in the book Penrose adds that

> Yet I hold also to the hope that it is through science and mathematics that some profound advances in the understanding of mind must eventually come to light.

Paul Davies (2020) put it that, when asked

> whether physics can explain life, most physicists would answer yes. The
> more pertinent question, however, is whether known physics is up to
> the job, or whether something fundamentally new is required.

We have a good start but many open questions, ample issues left for
research in the natural sciences to fascinate generations to come.

## 7.5  *The Future*

Might the history of research in physical science offer hints to what
the future will bring? Cosmology gives us a clear picture of how a
science grew, a picture biased by its simplicity but still a worked
example to instruct us. Cosmology in the year 1940 was an inviting shell of ideas nearly free of empirical considerations. Research
in the cosmology of 1970 included the analyses of a rich variety of
physical processes, albeit more often than not applied in speculative
ways because of the improving but still scant empirical constraints
on our imaginations. Cosmology in the year 2000 had grown into
a big science with large research groups working on well-defined
experimental goals that the community supported because they
were judged to be well motivated. Small groups had a significant
place too, as sources for adventurous ideas about the nature of the
dark sector of the $\Lambda$CDM cosmology, and for ideas about what the
universe might have been doing before the big bang. Observations
by small groups also were adding to constraints on the present state
of the universe, as in the close examination of individual nearby
galaxies and the intergalactic matter around them. This was hand-
work for the baseline essential for comparison to what was inferred
from the evidence gathered by big science on the nature of the
universe at high redshift, the universe as it used to be.

So what might be the balance of research in physical cosmology,
directed or speculative, in groups large or small, in 2030, a decade
from the time of writing? An indicator will be the state of comparisons of predictions and measurements of parameters, along the
lines of the table in Section 6.10. If the measurement uncertainties are smaller, and there remain tantalizing but not compelling
indications of inconsistency with the predictions of the $\Lambda$CDM

theory, we may expect that large groups will be working on big science projects aimed at improving the precision and accuracy of the empirical tests of the theory. If in 2030 such tests have already shown clear inconsistency with the predictions of ΛCDM, and no ready adjustment of the theory remedies the situation, then we may expect an intense search for new ideas, as always best done in small groups. Theorists are inventive; there will be promising-looking ideas. Observers will be occupied checking them and searching for hints to where to go next. If a better theory that removes all the present inconsistencies in physical cosmology has been found by 2030, then I expect two things. First, predictions of the new improved theory will be quite similar to ΛCDM, because this theory already passes demanding tests. That is, I expect the new theory will have a lot in common with ΛCDM. Second, I expect there will not be a basis for the pronouncement of a final theory. Experience suggests that a new theory that better fits the improved data of 2030 will be accompanied by new concerns about more subtle issues of incompleteness and inconsistency. This is the situation Poincaré (1902) envisioned: advances in research in successive approximations all the way down. Since physical cosmology depends on complex physics such as star formation, and on the data-starved exploration of what the universe was doing before it was expanding, I expect the cycle of search and discovery in physical cosmology to end in exhaustion rather than the claim of empirical establishment of a final theory.

This story of a century of development of physical cosmology is not a template for search and discovery in other branches of natural science, because conditions differ. But it is a useful existence proof for the viability of our scientific method.

## 7.6 On Reality

Though the power of science is clear from its practical applications, and has been for a long time, we have seen two directions of thought about what this power might mean. Is research in the natural sciences producing useful approximations to objective physical reality? Or is it constructing pictures of selected aspects of a world that is so deeply complex that the notion of objective reality is meaningless? Are the facts claimed by scientists reliable approximations to reality, established because they have been found to be about the same no matter who chooses to look into them? Or are the scientists' facts no more than social constructions built out of bits and pieces chosen for convenience from the bewildering world around us, and interpreted in a manner that is determined by the cultural history of the people who are building the story? Those who take an interest in the nature and meaning of what people working in the natural sciences have been doing were divided on this issue a century ago, and they still are. You can read about the subtleties of such issues from philosophers' points of view in the Stanford Encyclopedia of Philosophy entry on Models in Science (Frigg and Hartmann 2020). I offer my view from the physical scientists' side.

I see four points that favor the side of social constructionism. First, scientists keep changing their stories by adding paradigms. The picture of our Milky Way galaxy of stars changed a quarter of a century ago by the addition of a massive halo of nonbaryonic dark matter. So now the theory has it that this dark matter is coursing through your body. How can you trust a story that keeps changing, and on the face of it seems implausible? Second, while natural scientists are offended by those who question the scientists' notion of reality, and their suggestion that our theories are social constructions, natural scientists routinely use what clearly are social constructions. Examples are the general theory of relativity and the big bang theory of the expanding universe in 1960, the hypothetical nonbaryonic dark matter in 1990, and the concept of cosmological inflation now. The progress of research has resulted in promotion of all but the last of these examples to empirical constructions: more changes to the scientists' story. Third, the world around us is indeed wonderfully complex. That makes it easy to ask simple questions that scientists cannot answer. The somatostatin and thyrotropin-releasing hormones figure in Bruno Latour's observations while embedded as an anthropological observer in the Salk Institute for Biological Studies. If the physicists' theories are so good why can't physicists account for the properties of these hormones? Susan Haack's (2009) assessment of the situation is that there is no reason to think that it [science] is in possession of a special method of inquiry unavailable to historians or detectives or the rest of us, nor that it is immune from the susceptibility to fad and fashion, politics and propaganda, partiality and power-seeking to which all human cognitive activity is prone.... But *science* has had notable success

And the fourth point: what am I to make of the phenomenon, as I think I may put it, of being conscious and forming opinions about what it all means? Might the deep complexity of this phenomenon make the idea of objective reality meaningless? I have nothing more to add to many generations of thought.

Yet another point mentioned in Section 2.1 in the Stanford Encyclopedia of Philosophy entry on Scientific Realism (Chakravartty 2017) is that

since there is no independent way of knowing the base rate of approximately true theories, the chances of it being approximately true cannot be assessed.

Although we celebrate the broad and well-checked tests that the ΛCDM theory passes we cannot demonstrate that there is not a different theory of the large-scale nature of our universe that passes its own broad array of tests. We must resort to intuition: the idea seems wildly unlikely to me.

I open the argument in defense of objective reality by pointing to the experience of facts. I expect shipbuilders in Venice ten centuries ago were quite aware of the fact that a newly constructed ship either is stable or else unstable and liable to turn over and sink. The latter would have been a fact for all who were there to see. Six centuries ago Galileo observed the regularity of the period of a pendulum given its length. This was not a construct forced on people by an elite class, it was a fact that Galileo and others could demonstrate, on demand. In Steven Weinberg's (1996) example, "when we stub our toe on an unnoticed rock" we can be sure that witnesses who inspect the scene will observe the rock, the fact on the ground. Newton's theory of the motions of the planets and their moons could be checked by data from the long history of observations of where the moons and planets have been, data obtained by curiosity-driven observers who had other things on their minds than the prospect of some future theory. The consistency of these old observations with the Newtonian theory when it was introduced is a fact on the ground.

There is ample room in these examples for Haack's "susceptibility to fad and fashion"; we all are prone to it. But could we imagine that the reported consistency of Newtonian physics with the older observations, and the consistency with subsequent ones (apart from tiny relativistic corrections), is a fable? It looks so unlikely. The consistency instead looks like Haack's "notable success."

The measurements shown in Figure 6.1 cannot be checked against earlier evidence from disinterested observers, but the careful checks of reproducibility by comparisons of measurements from different ground-based instruments and the two satellite missions, WMAP and PLANCK, make a persuasive case that the data shown in the figure are reproducible, always within measurement uncertainties of course.

The demonstration of consistency of the $\Lambda$CDM theory with the well-checked measurements in Figure 6.1 was contingent on someone hitting on the theory. The multiple paths to this theory from the empirical clues reviewed in Section 7.1 make this contingency seem close to an inevitability. And the clear reading of the evidence is that the consistency of the $\Lambda$CDM theory with the observations reviewed in Section 6.10 is another "notable success" of physical science.

The search for better physics, what Mach considered its "disproportionate formal development," gave us the electromagnetic field more than a century ago, then atoms and electrons, then the state vectors of quantum physics, and by the end of the last century a standard theory for the large-scale nature and evolution of the universe. These increasingly detailed, one might say increasingly formal, theories were constructed by elite groups that society has been willing to support, but there certainly have been enough of them who have been quite willing and able to check on each other to establish a compelling case that the successes of these theories are more facts on the ground. This is never a proof of course, but it is difficult to ignore if you do not wish to deal with a considerable number of Putnam's miracles.

The scientific method in the search for facts on the ground depends on checks for reproducibility of experiments and observations when not disturbed by outside interference, as by a prankster who removes the rock Weinberg stumbled on before witnesses could inspect the site. I mentioned (on pages 7, 82, and 171) the concern bordering on obsession with systematic errors caused by undetected disturbances. Control of systematic errors is essential to the scientist's concern with checks of reproducibility. The checks—repeat the measurements, let others repeat them, maybe by different methods—are in turn an essential probe for systematic errors that might be different in different ways to make the measurement. It is human nature that the search for systematic errors seems less urgent when the results agree with what is expected; it happens. Concern about this leads to blind tests: parallel measurements by independent groups who keep their results private until all have definite results to compare. You see where I am heading: we have a persuasive case for facts on the ground from the checks of reproducibility and tests for systematic errors. The results speak of a reality that operates in what we would consider a rational way.

Natural scientists are not consistently rational. We keep changing our stories by adding paradigms and promoting and demoting social and empirical constructions. When I started looking into the relativistic big bang cosmology in the early 1960s I was dismayed by the close to non-existent empirical support. In those days the status of cosmology was not even what I have learned to term a social construction, because not many in the community were confident that cosmology was a "real science." I kept looking into cosmology because there were many interesting physical considerations to analyze and compare to whatever empirical constraints I could find, the kind of physics I enjoy doing. But non-specialists who observed me then could see sensible reasons to wonder whether they were seeing the constructions of just so stories. Explorations of adventurous ideas are part of the productive search for more complete physics, but of course ideas must not be confused with empirically tested facts. Adventurous ideas such as the multiverse may be elegant, newsworthy, constructive, distracting, or wrong; recall the elegance of the 1948 steady-state cosmology. But adventurous ideas have aided discovery of facts of natural science, and scientists are not about to give up exploring them.

Adventurous theories are sorted out, and some promoted to facts, by empirical checks of predictions, another of our obsessions. A century ago Peirce (1878a) presented a good example of how this goes. Electromagnetic theory was changing the way society operates: electric lights, street cars, communication by telegraph and radio. This theory predicts a value of the speed of light that is consistent with what was measured in quite different ways by applications of Newtonian physics to observations of transits of Venus, oppositions of Mars, and eclipses of Jupiter's moons. These were results of curiosity-driven research in astronomy and the laboratory; the speed of light was a by-product of the primary research interests of most of the people doing the observations and experiments. It would be absurd to imagine these people somehow conspired to shape choices of data and analyses to arrive at the same value of the speed of light. It seems equally absurd to imagine the consistency of measures of the speed of light is one of Putnam's miracles. The consistency presents us with an impressive case for the predictive power of physical theories.

A more recent illustration of this predictive power is the figure on page 173, which shows statistical measures of the patterns in the angular distribution of the fossil radiation and the spatial distribution of the galaxies. The interpretation is based on the physics we had in place by 1970 and the model we had by 1984, well before the measurements. It needn't have been that way, but that is what happened. In effect, the theory and model told the observers what to look for, and that is what they found.

The table on page 175 compares values of the parameters derived from the fit of the $\Lambda$CDM theory to a broad variety of measurements: statistics of the large-scale distributions of matter and radiation, measurements of the abundances of the lightest isotopes, observations of supernovae, and the dispersion measures of fast radio bursts. The measurements are interpreted by what is predicted to have been happening recently, at redshifts near zero; back in time to redshift $z \sim 1$; then to $z \sim 10^3$; and then really far back in time to $z \sim 10^9$. The consistency of theory and observation is impressive. I include the comparison of the two measures of Hubble's constant discussed on page 181. They are based on what is observed now and on what is inferred from the patterns set in the distributions of matter and radiation at decoupling, at redshift $z \sim 10^3$. At the time of writing, the two measures differ by about five to ten percent. If confirmed the difference would be exciting, a hint to how to improve our theory of cosmic evolution. But the important point for the present discussion is that the discrepancy is ten percent or so based on what is happening now and what was happening when the mean distance between particles was a thousandth of the present value. This is impressive precision. Also impressive is the similar degree of consistency of independent measures of the other parameters in Table 6.1, based on what was happening at quite different epochs in the course of evolution of the universe.

The observations of the distributions of the galaxies and the fossil radiation used for Figure 6.1 were designed to test the prediction of a specific model, apart from the minor adjustment for tilt. One might object that we found what we were looking for. But it is really difficult to imagine that the groups measuring these distributions unintentionally conspired to arrive at consistency; we do not operate that way. Recall the excitement occasioned by the possibility of a ten percent difference in the value of Hubble's constant obtained from two different sets of considerations. Consider the excitement that would be occasioned by the discovery that the parameters in the $\Lambda$CDM theory that fit the radiation measurements in Panel (a) of Figure 6.1 do not quite fit the matter measurements in Panel (b). Again you see where I am heading. The evidence on the ground of the quite close consistency of predictions from different theories and observations agrees with the assumption that reality operates in what we would consider a rational and lawful way, and with the observation that our theories appear to be good enough approximations to this assumed reality that they predict phenomena beyond those used to find the theories.

The arguments from physical science for objective reality are based on situations that are simple, relatively speaking, and allow the tidy tests we have for the fundamental theories of relativity and quantum physics. Are the properties of living systems based on the same fundamental physics, or maybe on an even better theory that passes the tests we have now and tighter ones to be discovered? The empirical argument that this is likely to be so is the tight reproducibility of experiments in biological sciences, a fact on the ground. The theoretical argument is that if the simpler examples in physical science are accepted as approximations to reality, which is hard to resist, it certainly encourages the idea that complex objects operate on the same objective grounds. The complexity of these systems, in structure and environment, prevents the clear tests we have in physical science with the methods we have now. The search for tests of the fundamental basis for more complicated phenomena will continue to be explored by generations of scientists to come, assuming society continues to make it possible.

After all these considerations, how shall we frame a definition of reality in a manner that aids our intent to defend the idea? Peirce's (1878a) definition (discussed on page 5, presented here in a later version) is that

> The opinion which is <u>fated</u> to be ultimately agreed to by all who investigate, is what we mean by the truth, and the object represented in this opinion is the real. That is the way I would explain reality.

This is operational, which is good. It is pragmatic, as Peirce (1869, page 208) put it, because it is "justified by its indispensibleness for making any action rational." Peirce implicitly assumed that reality behaves in what we would say is a rational way, and that "all who investigate" will arrive at the same result. We have not encountered any evidence since then that causes the physical science community to reconsider this assumption.

I mentioned the thought that the empirical evidence that guides our thinking might offer convergence to more than one basin of attraction. If those who investigate could arrive at two physically different theories, both of which pass the test of predictions of the evidence, it would falsify the assumption that we can comprehend reality (Hoyningen-Huene 2013). I see no indication that this is a threat to the ways of natural science, but we should be aware of such things.

The philosopher Bruno Latour and the sociologist Steve Woolgar (1979, quoted on page 33) proposed that "that which cannot be changed at will is what counts as real." With Peirce, this is operational, and with Mach, it overlooks the remarkable predictive power of physical theory when applied to situations where predictions and tests can be trusted. But Latour and Woolgar were not well positioned to draw this lesson from Latour's observations of research at the Salk Institute for Biological Studies, because the study of living matter is really complicated.

The natural science community finesses the challenge of defining objective physical reality by instead working with the assumption of reality that has properties we can discover from the many clues to be found by the study of the world around us, on scales large and small, and studied in so many different ways. The plan is not new; a century ago Peirce expressed similar thoughts that he illustrated by the impressive predictive power of electromagnetism and Newtonian physics. It is what is expected if reality operates by rules and if electromagnetism and Newtonian physics are useful approximations to the rules. The reach of natural science is far greater now, and the well-checked tests of predictions far broader. This great reach and success of predictability makes the prime case for objective reality.

The methods of discovery in natural science are many. Compare Mach's search for the economy of nature revealed by strictly empirical means to Einstein's search for what seemed to him to be intuitively right and proper theories. Gamow's idea that the early universe was hot was a brilliant intuitive guess; the large abundance of helium was a suggestive empirical clue. We have learned a lot by such methods, and we have many inviting prospects for still more to learn about the world around us and the nature of what certainly seems to be objective physical reality.

# REFERENCES

Abbott, T. M. C., Aguena, M., Alarcon, A., et al. 2020. Dark Energy Survey Year 1 Results: Cosmological constraints from cluster abundances and weak lensing. Physical Review D 102, 023509, 34 pp.

Abell, G. O. 1958. The Distribution of Rich Clusters of Galaxies. Astrophysical Journal, Supplement 3, 211–288

Adams, W. S. 1925. The Relativity Displacement of the Spectral Lines in the Companion of Sirius. Proceedings of the National Academy of Science 11, 382–387

Aller, L. H., and Menzel, D. H. 1945. Physical Processes in Gaseous Nebulae XVIII. The Chemical Composition of the Planetary Nebulae. Astrophysical Journal 102, 239–263

Alpher, R. A., Bethe, H., and Gamow, G. 1948. The Origin of Chemical Elements. Physical Review 73, 803–804

Alpher, R. A., Follin, J. W., and Herman, R. C.1953. Physical Conditions in the Initial Stages of the Expanding Universe. Physical Review 92, 1347–1361

Alpher, R. A., and Herman, R. C. 1948. Evolution of the Universe. Nature 162, 774–775

Alpher, R. A., and Herman, R. C. 1950. Theory of the Origin and Relative Abundance Distribution of the Elements. Reviews of Modern Physics 22, 153–212

Aver, E., Berg, D. A., Olive, K. A., et al. 2020. Improving Helium Abundance Determinations with Leo P as a Case Study. arXiv:2010.04180

Baade, W. 1939. Stellar Photography in the Red Region of the Spectrum. Publications of the American Astronomical Society 9, 31–32

Babcock, H. W. 1939. The Rotation of the Andromeda Nebula. Lick Observatory Bulletin 498, 41–51

Baghramian, M., and Carter, J. A. 2021. Relativism, in The Stanford Encyclopedia of Philosophy (Spring 2021 Edition), Edward N. Zalta (ed.) https://plato.stanford.edu/archives/spr2021/entries/relativism/

Bahcall, N. A., and Cen, R. 1992. Galaxy Clusters and Cold Dark Matter: A Low-Density Unbiased Universe? Astrophysical Journal Letters 398, L81–L84

Bahcall, N. A., Fan, X., and Cen, R. 1997. Constraining $\Omega$ with Cluster Evolution. Astrophysical Journal Letters 485, L53–L56

Bennett, A. S. 1962. The Preparation of the Revised 3C Catalogue of Radio Sources. Monthly Notices of the Royal Astronomical Society 125, 75–86

Bergmann, P. G. 1942. *Introduction to the Theory of Relativity*. New York: Prentice-Hall

Bertotti, B., Brill, D., and Krotkov, R. 1962. Experiments on Gravitation. In *Gravitation: An Introduction to Current Research*. Ed. L. Witten. Wiley, New York, pp. 1–48

Bertotti, B., Iess, L., and Tortora, P. 2003. A Test of General Relativity Using Radio Links with the Cassini Spacecraft. Nature 425, 374–376

Beutler, F., Seo, H.-J., Ross, A. J., et al. 2017. The Clustering of Galaxies in the Completed SDSS-III Baryon Oscillation Spectroscopic Survey: Baryon Acoustic Oscillations in the Fourier Space. Monthly Notices of the Royal Astronomical Society 464, 3409–3430

Bloor, D. 1976. *Knowledge and Social Imagery*. London: Routledge & Kegan Paul

Bloor, D. 1991. *Knowledge and Social Imagery*. Second Edition, Chicago: The University of Chicago Press

Bondi, H. 1952. *Cosmology*. Cambridge: Cambridge University Press

Bondi, H. 1960. *Cosmology*, second edition. Cambridge: Cambridge University Press

Bondi, H., and Gold, T. 1948. The Steady-State Theory of the Expanding Universe. Monthly Notices of the Royal Astronomical Society 108, 252–270

Boughn, S. P., Cheng, E. S., and Wilkinson, D. T. 1981. Dipole and Quadrupole Anisotropy of the 2.7 K Radiation. Astrophysical Journal Letters 243, L113–L117

Brans, C., and Dicke, R. H. 1961, Physical Review, 124, 925

Brault, J. W. 1962. The Gravitational Red Shift in the Solar Spectrum. Ph.D. Thesis, Princeton University

Brent, J. 1993. *Charles Sanders Peirce: A Life*. Bloomington: Indiana University Press

Burbidge, E. M., Burbidge, G. R., Fowler, W. A., and Hoyle, F. 1957. Synthesis of the Elements in Stars. Reviews of Modern Physics 29, 547–650

Burbidge, G. R. 1958. Nuclear Energy Generation and Dissipation in Galaxies. Publications of the Astronomical Society of the Pacific 70, 83–89

Burbidge, G. R. 1962. Nuclear Astrophysics. Annual Review of Nuclear and Particle Science 12, 507–572

Campbell, W. W. 1909. The Closing of a Famous Astronomical Problem. Publications of the Astronomical Society of the Pacific 21, 103–115

Campbell, W. W., and Trumpler, R. J. 1928. Observations Made with a Pair of Five-foot Cameras on the Light-deflections in the Sun's Gravitational Field at the Total Solar Eclipse of September 21, 1922. Lick Observatory Bulletin 397, 130–160

Carnap, R. 1963. In *The Philosophy of Rudolph Carnap*. Illinois: Open Court Publishing Company

Chakravartty, A. 2017. Scientific Realism, in The Stanford Encyclopedia of Philosophy (Summer 2017 Edition), Edward N. Zalta (ed.) https://plato.stanford.edu/archives/sum2017/entries/scientific-realism/

Chandrasekhar, S. 1935. Stellar Configurations with Degenerate Cores (Second paper). Monthly Notices of the Royal Astronomical Society 95, 676–693

Chandrasekhar, S., and Henrich, L. R. 1942. An Attempt to Interpret the Relative Abundances of the Elements and Their Isotopes. Astrophysical Journal 95, 288–298

Charlier, C. V. L. 1922. How an Infinite World May Be Built Up. Meddelanden fran Lunds Astronomiska Observatorium. Serie I 98, 1–37

Collins, H. M., and Pinch, T. J. 1993. *The Golems: What Everyone Should Know about Science*. Cambridge: Cambridge University Press

Comesaña, J., and Klein, P. 2019. Skepticism, in The Stanford Encyclopedia of Philosophy (Winter 2019 Edition), Edward N. Zalta (ed.) https://plato.stanford.edu/archives/win2019/entries/skepticism/

Comte, A. 1896. The Positive Philosophy of Auguste Comte, Volume 1, freely translated from French and condensed by Harriet Martineau. London: George Bell & Sons

Cooke, R. J., Pettini, M., and Steidel, C. C. 2018. One Percent Determination of the Primordial Deuterium Abundance. Astrophysical Journal 855:102, 16 pp.

Corry, L. 1999. From Mie's Electromagnetic Theory of Matter to Hilbert's Unified Foundations of Physics. Studies in History and Philosophy of Modern Physics 30B, 159–183

Corry, L., Renn, J., and Stachel, J. 1997. Belated Decision in the Hilbert-Einstein Priority Dispute. Science 278, 1270–1273

Cowsik, R., and McClelland, J. 1972. An Upper Limit on the Neutrino Rest Mass. Physical Review Letters 29, 669–670

Cowsik, R., and McClelland, J. 1973. Gravity of Neutrinos of Nonzero Mass in Astrophysics. Astrophysical Journal 180, 7–10

Davies, P. 2020. Does New Physics Lurk Inside Living Matter? Physics Today 73, 34–40

Davis, M., Groth, E. J., and Peebles, P. J. E. 1977. Study of Galaxy Correlations: Evidence for the Gravitational Instability Picture in a Dense Universe. Astrophysical Journal, Letters 212, L107–L111

Davis, M., Huchra, J., Latham, D. W., and Tonry, J. 1982. A Survey of Galaxy Redshifts. II. The Large Scale Space Distribution. Astrophysical Journal 253, 423–445

Davis, M., and Peebles, P. J. E. 1983. A Survey of Galaxy Redshifts. V. The Two-Point Position and Velocity Correlations. Astrophysical Journal 267, 465–482

Dawid, R. 2013. *String Theory and the Scientific Method.* Cambridge: Cambridge University Press

DeGrasse, R. W., Hogg, D. C., Ohm, E. A., and Scovil, H. E. D. 1959a. Ultra-Low-Noise Measurements Using a Horn Reflector Antenna and a Traveling-Wave Maser. Journal of Applied Physics 30, 2013

DeGrasse, R. W., Hogg, D. C., Ohm, E. A., and Scovil, H. E. D. 1959b. Ultra-Low-Noise Antenna and Receiver Combination for Satellite or Space Communication. Proceedings of the National Electronics Conference 15, 371–379

de Sitter, W. 1913. The Secular Variation of the Elements of the Four Inner Planets. The Observatory 36, 296–303

de Sitter, W. 1916. On Einstein's Theory of Gravitation and Its Astronomical Consequences. Second paper. Monthly Notices of the Royal Astronomical Society 77, 155–184

de Sitter, W. 1917. On the Relativity of Inertia. Remarks Concerning Einstein's Latest Hypothesis. Koninklijke Nederlandse Akademie van Wetenschappen Proceedings Series B Physical Sciences 19, 1217–1225 de Vaucouleurs, G. 1958. Photoelectric Photometry of the Andromeda Nebula in the UBV System. Astrophysical Journal 128, 465–488

de Vaucouleurs, G. 1970. The Case for a Hierarchical Cosmology. Science 167, 1203–1313

Dewey, J. 1903, *Studies in Logical Theory*, Chicago: The University of Chicago Press

Dicke, R. H. 1957. The Experimental Basis of Einstein's Theory. In Proceedings of the Conference on the Role of Gravitation in Physics at the University of North Carolina, Chapel Hill, January 18–23, 1957, pp. 5–12. Eds. Cécile DeWitt and Bryce DeWitt.

Dicke, R. H. 1961. Dirac's Cosmology and Mach's Principle. Nature 192, 440–441

Dicke, R. H. 1964. *The Theoretical Significance of Experimental Relativity*. New York: Gordon and Breach

Dicke, R. H. 1970. Gravitation and the Universe. Memoirs of the American Philosophical Society, Jayne Lectures for 1969. Philadelphia: American Philosophical Society

Dicke, R. H., Beringer, R., Kyhl, R. L., and Vane, A. B. 1946. Atmospheric Absorption Measurements with a Microwave Radiometer. Physical Review 70, 340–348

Dicke, R. H., and Peebles, P. J. E. 1965. Gravitation and Space Science. Space Science Reviews 4, 419–460

Dicke, R. H., and Peebles, P. J. E. 1979. The Big Bang Cosmology—Enigmas and Nostrums. In *General Relativity: An Einstein Centenary Survey*, pp. 504–517. Eds. S. W. Hawking and W. Israel. Cambridge: Cambridge University Press

Dicke, R. H., Peebles, P. J. E., Roll, P. G., and Wilkinson, D. T. 1965. Cosmic Black-Body Radiation. Astrophysical Journal 142, 414–419

Dicus, D. A., Kolb, E. W., and Teplitz, V. L. 1977. Cosmological Upper Bound on Heavy-Neutrino Lifetimes. Physical Review Letters 39, 168–171

Dingle, H. 1931. The Nature and Scope of Physical Science. Nature 127, 526–527

Di Valentino, E., Anchordoqui, L. A., Akarsu, O., et al. 2020. Cosmology Intertwined II: The Hubble Constant Tension. Snowmass 2021—Letter of Interest

Di Valentino, E., Anchordoqui, L. A., Akarsu, O., et al. 2020. Cosmology Intertwined IV: The Age of the Universe and Its Curvature. Snowmass 2021—Letter of Interest

Doroshkevich, A. G., and Novikov, I. D. 1964. Mean Density of Radiation in the Metagalaxy and Certain Problems in Relativistic Cosmology. Doklady Akademii Nauk SSSR 154, 809–811. English translation in Soviet Physics Doklady 9, 111–113

Dyson, F. W., Eddington, A. S., and Davidson, C. 1920. A Determination of the Deflection of Light by the Sun's Gravitational Field, from Observations Made at the Total Eclipse of May 29, 1919. Philosophical Transactions of the Royal Society of London Series A 220, 291–333

Earman, J., and Glymour, C. 1980a. The Gravitational Red Shift as a Test of General Relativity: History and Analysis. Studies in the History and Philosophy of Science 11, 175–214

Earman, J. and Glymour, C. 1980b. Relativity and Eclipses: The British Eclipse Expeditions of 1919 and Their Predecessors. Historical Studies in the Physical Sciences 11, 49–85

Eddington, A. S. 1914. *Stellar Movements and the Structure of the Universe*. London: Macmillan

Eddington, A. S., Jeans, J. H., Lodge, O., et al. 1919. Discussion on the Theory of Relativity. Monthly Notices of the Royal Astronomical Society 80, 96–118

Einstein, A. 1917. Kosmologische Betrachtungen zur allgemeinen Relativitätstheorie. Sitzungsberichte der Königlich Preußischen Akademie der Wissenschaften, Berlin, pp. 142–152

Einstein, A. 1922a. *The Meaning of Relativity*. Princeton: Princeton University Press

Einstein, A. 1922b. *Sidelights on Relativity*. London: Methuen

Einstein, A. 1936. Physics and Reality. Journal of The Franklin Institute 221, 349–382

Einstein, A., and de Sitter, W. 1932. On the Relation Between the Expansion and the Mean Density of the Universe. Proceedings of the National Academy of Science 18, 213–214

Enz, C. P., and Thellung, A. 1960. Nullpunktsenergie und Anordnung nicht vertauschbarer Faktoren im Hamiltonoperator. Helvetica Physica Acta 33, 839–848

Fabbri, R., Guidi, I., Melchiorri, F., and Natale, V. 1980. Measurement of the Cosmic-Background Large-Scale Anisotropy in the Millimetric Region. Physical Review Letters 44, 1563–1566

Fermi, E. 1934. Versuch einer Theorie der $\beta$-Strahlen. Zeitschrift für Physik 88, 161–177

Feynman, R. P., Morinigo, F. B., and Wagner, W. G. 1995. *Feynman Lectures on Gravitation*. Reading, MA: Addison-Wesley

Finlay-Freundlich, E. 1954. Red Shifts in the Spectra of Celestial Bodies. The London, Edinburgh, and Dublin Philosophical Magazine and Journal of Science 45, 303–319

Fixsen, D. J., Cheng, E. S., and Wilkinson, D. T. 1983. Large-Scale Anisotropy in the 2.7-K Radiation with a Balloon-Borne Maser Radiometer at 24.5 GHz. Physical Review Letters 50, 620–622

Frank, P. 1949. *Modern Science and Its Philosophy*. Cambridge: Harvard University Press

Franklin, A. 2005. *No Easy Answers: Science and the Pursuit of Knowledge*. Pittsburgh: University of Pittsburgh Press

Freeman, K. C. 1970. On the Disks of Spiral and S0 Galaxies. Astrophysical Journal 160, 811–830

Friedman, A. 1922. Über die Krümmung des Raumes. Zeitschrift für Physik 10, 377–386

Friedman, A. 1924. Über die Möglichkeit einer Welt mit Konstanter Negativer Krümmung des Raumes. Zeitschrift für Physik 21, 326–332

Frigg, R., and Hartmann, S. 2020. Models in Science, in The Stanford Encyclopedia of Philosophy (Spring 2020 Edition), Edward N. Zalta (ed.) https://plato.stanford.edu/archives/spr2020/entries/models-science/

Galison, P. 2016. Practice All the Way Down. In Kuhn's *Structure of Scientific Revolutions* at Fifty: Reflections on a Science Classic, pp. 42–70. Eds. Robert J. Richards and Lorraine Daston. Chicago: University of Chicago Press

Gamow, G. 1937. *Structure of Atomic Nuclei and Nuclear Transformations*. Oxford: The Clarendon Press

Gamow, G. 1946. Expanding Universe and the Origin of Elements. Physical Review 70, 572–573

Gamow, G. 1948a. The Origin of Elements and the Separation of Galaxies. Physical Review 74, 505–506

Gamow, G. 1948b. The Evolution of the Universe. Nature 162, 680–682

Gamow, G. 1949. On Relativistic Cosmogony. Reviews of Modern Physics 21, 367–373

Gamow, G. 1953a. In Proceedings of the Michigan Symposium on Astrophysics, June 29-July 24, 1953, 29 pp.

Gamow, G. 1953b. Expanding Universe and the Origin of Galaxies. Danske Matematisk-fysiske Meddelelser 27, number 10, 15 pp.

Gamow, G., and Critchfield, C. L. 1949. *Theory of Atomic Nucleus and Nuclear Energy-Sources*. Oxford: Clarenden Press

Gamow, G., and Teller, E. 1939. The Expanding Universe and the Origin of the Great Nebulæ. Nature 143, 116–117

Gershtein, S. S., and Zel'dovich, Y. B. 1966. Rest Mass of Muonic Neutrino and Cosmology. Zhurnal Eksperimental'noi i Teoreticheskoi Fiziki Pis'ma 4, 174–177. English translation in Journal of Experimental and Theoretical Physics Letters 4, 120–122

Greenstein, J. L., Oke, J. B., and Shipman, H. L. 1971. Effective Temperature, Radius, and Gravitational Redshift of Sirius B. Astrophysical Journal 169, 563–566

Greenstein, J. L., Oke, J. B., and Shipman, H. L. 1985. On the Redshift of Sirius B. Quarterly Journal of the Royal Astronomical Society 26, 279–288

Greenstein, J. L., and Trimble, V. 1972. The Gravitational Redshift of 40 Eridani B. Astrophysical Journal, Letters 175, L1–L5

Groth, E. J., and Peebles, P. J. E. 1977. Statistical Analysis of Catalogs of Extragalactic Objects. VII. Two- and Three-Point Correlation Functions for the High-Resolution Shane-Wirtanen Catalog of Galaxies. Astrophysical Journal 217, 385–405

Gunn, J. E., Lee, B. W., Lerche, I., Schramm, D. N., and Steigman, G. 1978. Some Astrophysical Consequences of the Existence of a Heavy Stable Neutral Lepton. Astrophysical Journal 223, 1015–1031

Gunn, J. E., and Tinsley, B. M. 1975. An Accelerating Universe. Nature 257, 454–457

Gush, H. P. 1974. An Attempt to Measure the Far Infrared Spectrum of the Cosmic Background Radiation. Canadian Journal of Physics 52, 554–561

Gush, H. P., Halpern, M., and Wishnow, E. H. 1990. Rocket Measurement of the Cosmic-Background-Radiation mm-Wave Spectrum. Physical Review Letters 65, 537–540

Gutfreund, H., and Renn, J. 2017. *The Formative Years of Relativity: The History and Meaning of Einstein's Princeton Lectures*. Princeton: Princeton University Press

Guth, A. H. 1981. Inflationary Universe: A Possible Solution to the Horizon and Flatness Problems. Physical Review D 23, 347–356

Haack, S. 2009. *Evidence and Inquiry: A Pragmatist Reconstruction of Epistemology*, second expanded edition. Amherst, New York: Prometheus Books

Hacking, I. 1983. *Representing and Intervening: Introductory Topics in the Philosophy of Natural Science*. Cambridge: Cambridge University Press

Hacking, I. 1999. *The Social Construction of What?* Cambridge: Harvard University Press

Hahn, H, Neurath, O., and Carnap, R. 1929. Wissenschaftliche Weltauffassung Der Wiener Kreis. Vienna: Artur Wolf Verlag

Halsted, G. B. 1913. *The Foundations of Science: Science and Hypothesis, the Value of Science, Science and Method*. New York: The Science Press. http://www.gutenberg.org/ebooks/39713

Hang, Q., Alam, S., Peacock, J. A., et al. 2021. Galaxy Clustering in the DESI Legacy Survey and Its Imprint on the CMB. Monthly Notices of the Royal Astronomical Society, 501, 1481–1498

Harman, G. *Thought*. Princeton, NJ : Princeton University Press

Harrison, E. R. 1970. Fluctuations at the Threshold of Classical Cosmology. Physical Review D 1, 2726–2730

Hartshorne, C., and Weiss, P. 1934. The Collected Papers of Charles Sanders Peirce. Cambridge: Harvard University Press

Harvey, G. M. 1979. Gravitational Deflection of Light. The Observatory 99, 195–198

Herzberg, G. 1950. *Molecular Spectra and Molecular Structure*. I. Spectra of Diatomic Molecules, second edition. New York: Van Nostrand

Hetherington, N. S. 1980. Sirius B and the Gravitational Redshift: An Historical Review. Quarterly Journal of the Royal Astronomical Society 21, 246–252

Hilbert, D. 1900. Mathematische Probleme. Göttinger Nachrichten, 253–297. English translation in Mathematical Problems. Bulletin of the American Mathematical Society, 8, 437–479, 1902

Hohl, F. 1970. Dynamical Evolution of Disk Galaxies. NASA Technical Report, NASA-TR R-343, 108 pp.

Holton, G. 1988. *Thematic Origins of Scientific Thought: Kepler to Einstein*. Revised second edition. Cambridge: Harvard University Press

Hoyle, F. 1948. A New Model for the Expanding Universe. Monthly Notices of the Royal Astronomical Society 108, 372–382

Hoyle, F. 1949. Stellar Evolution and the Expanding Universe. Nature 163, 196–198

Hoyle, F. 1950. Nuclear Energy. The Observatory 70, 194–195

Hoyle, F. 1958. The Astrophysical Implications of Element Synthesis. In Proceedings of a Conference at the Vatican Observatory, Castel Gandolfo, May 20–28, 1957, edited by D.J.K. O'Connell. Amsterdam: North-Holland, pp. 279–284

Hoyle, F., 1981. The Big Bang in Astronomy. New Scientist 19 November, 521–527

Hoyle, F., and Tayler, R. J. 1964. The Mystery of the Cosmic Helium Abundance. Nature 203, 1108–1110

Hoyle, F., and Wickramasinghe, N. C. 1988. Metallic Particles in Astronomy. Astrophysics and Space Science 147, 245–256

Hoyningen-Huene, P. 2013. The Ultimate Argument against Convergent Realism and Structural Realism: The Impasse Objection. In *Perspectives and Foundational Problems in Philosophy of Science*, Eds. V. Karakostas and D. Dieks, Springer International Publishing Switzerland, 131–139

Hubble, E. 1929. A Relation between Distance and Radial Velocity among Extra-Galactic Nebulae. Proceedings of the National Academy of Science 15, 168–173

Hubble, E. 1934. The Distribution of Extra-Galactic Nebulae. Astrophysical Journal 79, 8–76

Hubble, E, 1936. *The Realm of the Nebulae*. New Haven: Yale University Press

Hubble, E., and Humason, M. L. 1931. The Velocity-Distance Relation among Extra-Galactic Nebulae. Astrophysical Journal 74, 43–80

Hut, P. 1977. Limits on Masses and Number of Neutral Weakly Interacting Particles. Physics Letters B 69, 85–88

Huterer, D., and Turner, M. S. 1999. Prospects for Probing the Dark Energy via Supernova Distance Measurements. Physical Review D 60, 081301, 5 pp.

James, W. 1890. *The Principles of Psychology*. New York: Henry Holt and Company

James, W. 1907. *Pragmatism: A New Name for Some Old Ways of Thinking*. New York: Longmans, Green and Company

Janssen, M. 2014. No Success Like Failure: Einstein's Quest for General Relativity, 1907-1920. In *The Cambridge Companion to Einstein*, Eds. Michel Janssen and Cristoph Lehner. Cambridge: Cambridge University Press

Janssen, M., and Renn, J. 2015. Arch and Scaffold: How Einstein Found his Field Equations. Physics Today (November) 68, 30–36

Jeans, J. H. 1902. The Stability of a Spherical Nebula. Philosophical Transactions of the Royal Society of London Series A 199, 1–53

Jeans, J. H. 1928. *Astronomy and Cosmogony*. Cambridge: The University Press

Jeffreys, H. 1916. The Secular Perturbations of the Four Inner Planets. Monthly Notices of the Royal Astronomical Society 77, 112–118

Jeffreys, H. 1919. On the Crucial Tests of Einstein's Theory of Gravitation. Monthly Notices of the Royal Astronomical Society 80, 138–154

Jordan, P., and Pauli, W. 1928. Zur Quantenelektrodynamik ladungsfreier Felder. Zeitschrift fur Physik 47, 151–173

Kaluza, T. 1921. Zum Unitätsproblem der Physik. Sitzungsberichte der Preußischen Akademie der Wissenschaften zu Berlin 34, 966–972

Kelvin, L. 1901. Nineteenth Century Clouds over the Dynamical Theory of Heat and Light. Philosophical Magazine Series 6, 2:7, pp. 1–40

Kennefick, D. 2009. Testing Relativity from the 1919 Eclipse—A Question of Bias. Physics Today 62, 37–42

Kipling, R. 1902. *Just So Stories for Little Children*. New York: Doubleday

Klein, O. 1926. The Atomicity of Electricity as a Quantum Theory Law. Nature 118, 516

Klein, O. 1956. On the Eddington Relations and Their Possible Bearing on an Early State of the System of Galaxies. Helvetica Physica Acta Supplementum IV, 147–149

Knop, R. A., Aldering, G., Amanullah, R., et al. 2003. New Constraints on $\Omega_m$, $\Omega_\Lambda$, and $w$ from an Independent Set of 11 High-Redshift Supernovae Observed with the Hubble Space Telescope. Astrophysical Journal 598, 102–137

Knorr-Cetina, K. D. 1981. *The Manufacture of Knowledge: An Essay on the Constructivist and Contextual Nature of Science*. Elmsford, New York: Pergamon Press

Kuhn, T. S. 1970. *The Structure of Scientific Revolutions*. Chicago: University of Chicago Press

Kuhn, T. S., and van Vleck, J. H. 1950. A Simplified Method of Computing the Cohesive Energies of Monovalent Metals. Physical Review 79, 382–388

Kuiper, G. P. 1941. White Dwarfs: Discovery, Observations, Surface Conditions. Actualités Scientifiques et Industrielles. Paris: Hermann, pp. III-3 – III-39

Labinger, J. A., and Collins, H. 2001. *The One Culture? A Conversation about Science*. Chicago: University of Chicago Press

Ladyman, James 2020. Structural Realism, in The Stanford Encyclopedia of Philosophy (Winter 2020 Edition), Edward N. Zalta (ed.) https://plato.stanford.edu /archives/win2020/entries/structural-realism/

Landau, L., and Lifshitz, E. M. 1951. *The Classical Theory of Fields*. English translation of the 1948 second Russian edition. Cambridge USA: Addison-Wesley

Latour, B., and Woolgar, S. 1979. *Laboratory Life: The Social Construction of Scientific Facts*. Beverly Hills: Sage Publications

Latour, B., and Woolgar, S. 1986. *Laboratory Life: The Construction of Scientific Facts*. Princeton N.J.: Princeton University Press

Lee, B. W., and Weinberg, S. 1977. Cosmological Lower Bound on Heavy-Neutrino Masses. Physical Review Letters 39, 165–168

Lemaître, G. 1927. Un Univers homogène de masse constante et de rayon croissant rendant compte de la vitesse radiale des nébuleuses extra-galactiques. Annales de la Socíté Scientifique de Bruxelles A47, 49–59

Lemaître, G. 1931. The Expanding Universe. Monthly Notices of the Royal Astronomical Society 91, 490–501

Lemaître, G. 1934. Evolution of the Expanding Universe. Proceedings of the National Academy of Science 20, 12–17

Mach, E. 1883. *Die Mechanik in ihrer Entwicklung/historisch-kritisch dargestellt*. Leipzig: F. A. Brockhaus

Mach, E. 1898. *Popular Science Lectures*. Translated by Thomas J. McCormack. Chicago: The Open Court Publishing Company

Mach, E. 1902. *The Science of Mechanics: A Critical and Historical Account of Its Development*. Second revised and enlarged edition of the English translation. Chicago: The Open Court Publishing Company

Mach, E. 1960. *The Science of Mechanics: A Critical and Historical Account of Its Development*. Sixth edition of the English translation with revisions through the ninth German edition. Chicago: The Open Court Publishing Company

Macquart, J.-P., Prochaska, J. X., McQuinn, M., et al. 2020. A Census of Baryons in the Universe from Localized Fast Radio Bursts. Nature 581, 391–395

Mandelbrot, B. 1975. *Les objets fractals: Forme, hasard, et dimension*. Paris: Flammarion

Mandelbrot, B. 1989. *Les objets fractals: Survol du langage fractal*, third edition. Paris: Flammarion

Marx, G., and Szalay, A. S. 1972. Cosmological Limit on Neutretto Mass. In *Neutrino '72*, Balatonfüred Hungary, June 1972. Eds. A. Frenkel and G. Marx. Budapest: OMKDT-Technoinform I, 191–195

Mather, J. C., Cheng, E. S., Eplee, R. E., Jr., et al. 1990. A Preliminary Measurement of the Cosmic Microwave Background Spectrum by the Cosmic Background Explorer (COBE) Satellite. Astrophysical Journal Letters 354, L37–L40

Mayall, N. U. 1951. Comparison of Rotational Motions Observed in the Spirals M31 and M33 and in The Galaxy. Publications of the Observatory of the University of Michigan 10, 19–24

McCrea, W. H. 1971. The Cosmical Constant. Quarterly Journal of the Royal Astronomical Society 12, 140–153

McKellar, A. 1941. Molecular Lines from the Lowest States of Diatomic Molecules Composed of Atoms Probably Present in Interstellar Space. Publications of the Dominion Astrophysical Observatory 7, 251–272

McVittie, G. C. 1956. *General Relativity and Cosmology*. London: Chapman and Hall

Melott, A. L., Einasto, J., Saar, E., et al. 1983. Cluster Analysis of the Nonlinear Evolution of Large-Scale Structure in an Axion/Gravitino/Photino-Dominated Universe. Physical Review Letters 51, 935–938

Menand, L. 2001. *The Metaphysical Club: A Story of Ideas in America*. New York: Farrar, Straus and Giroux

Merton, R. K. 1961. Singletons and Multiples in Scientific Discovery: A Chapter in the Sociology of Science. Proceedings of the American Philosophy Society 105, 470–486

Merton, R. K. 1973. *The Sociology of Science: Theoretical and Empirical Investigations*. Chicago: The University of Chicago Press

Milgrom, M. 1983. A Modification of the Newtonian Dynamics as a Possible Alternative to the Hidden Mass Hypothesis. Astrophysical Journal 270, 365–370

Milne, E. A. 1932. World Structure and the Expansion of the Universe. Nature 130, 9–10

Milne, E. A. 1933. World-Structure and the Expansion of the Universe. Zeitschrift für Astrophysik 6, 1–95

Misak, C. 2013. *The American Pragmatists*. Oxford: Oxford University Press

Misak, C. 2016. *Cambridge Pragmatism: From Peirce and James to Ramsey and Wittgenstein*. Oxford: Oxford University Press

Mitton, S. 2005. *Fred Hoyle: A Life in Science*. London: Aurum Press

Møller, C. 1952. *The Theory of Relativity*. Oxford: Clarendon Press

Møller, C. 1957. On the Possibility of Terrestrial Tests of the General Theory of Relativity. Il Nuovo Cimento 6, 381–398

Moore, J. H. 1928. Recent Spectrographic Observations of the Companion of Sirius. Publications of the Astronomical Society of the Pacific 40, 229–233

Mössbauer, R. L. 1958. Kernresonanzfluoreszenz von Gammastrahlung in $Ir^{191}$. Zeitschrift für Physik 151, 124–143

Muhleman, D. O., Ekers, R. D., and Fomalont, E. B. 1970. Radio Interferometric Test of the General Relativistic Light Bending Near the Sun. Physical Review Letters 24, 1377–1380

O'Dell, C. R., Peimbert, M., and Kinman, T. D. 1964. The Planetary Nebula in the Globular Cluster M15. Astrophysical Journal 140, 119–129

Ogburn, W. F., and Thomas, D. S. 1922. Are Inventions Inevitable? A Note on Social Evolution. Political Science Quarterly 37, 83–98

Oort, J. H. 1958. Distribution of Galaxies and the Density of the Universe. Eleventh Solvay Conference. Brussels: Editions Stoops, pp. 163–181

Osterbrock, D. E. 2009. The Helium Content of the Universe. In *Finding the Big Bang*, Peebles, Page, and Partridge, pp. 86–92. Cambridge: Cambridge University Press

Osterbrock, D. E., and Rogerson, J. B., Jr. 1961. The Helium and Heavy-Element Content of Gaseous-Nebulae and the Sun. Publications of the Astronomical Society of the Pacific 73, 129–134

Ostriker, J. P., and Cowie, L. L. 1981. Galaxy Formation in an Intergalactic Medium Dominated by Explosions. Astrophysical Journal Letters 243, L127–L131

Ostriker, J. P., and Peebles, P. J. E. 1973. A Numerical Study of the Stability of Flattened Galaxies: or, Can Cold Galaxies Survive? Astrophysical Journal 186, 467–480

Pauli, W. 1933. Die allgemeinen Prinzipien der Wellenmechanik. Handbuch der Physik, Quantentheorie XXIV, part one (second edition), 83–272. Berlin: Springer

Pauli, W. 1958. *Theory of Relativity*. London: Pergamon Press

Peebles, P. J. E. 1964. The Structure and Composition of Jupiter and Saturn. Astrophysical Journal 140, 328–347

Peebles, P. J. E. 1971. *Physical Cosmology*. Princeton: Princeton University Press

Peebles, P. J. E. 1975. Statistical Analysis of Catalogs of Extragalactic Objects. VI. The Galaxy Distribution in the Jagellonian Field. Astrophysical Journal 196, 647–651

Peebles, P. J. E. 1980. *Large-Scale Structure of the Universe*. Princeton: Princeton University Press

Peebles, P. J. E. 1981. Large-Scale Fluctuations in the Microwave Background and the Small-Scale Clustering of Galaxies. Astrophysical Journal Letters 243, L119–L122

Peebles, P. J. E. 1982. Large-scale Background Temperature and Mass Fluctuations due to Scale-Invariant Primeval Perturbations. Astrophysical Journal Letters 263, L1–L5

Peebles, P. J. E. 1983. The Sequence of Cosmogony and the Nature of Primeval Departures from Homogeneity. Astrophysical Journal 274, 1–6

Peebles, P. J. E. 1984a. Dark Matter and the Origin of Galaxies and Globular Star Clusters. Astrophysical Journal 277, 470–477

Peebles, P. J. E. 1984b. Tests of Cosmological Models Constrained by Inflation. Astrophysical Journal 284, 439–444

Peebles, P. J. E. 1986. The Mean Mass Density of the Universe. Nature 321, 27–32

Peebles, P. J. E. 1987a. Origin of the Large-Scale Galaxy Peculiar Velocity Field: a Minimal Isocurvature Model. Nature 327, 210–211

Peebles, P. J. E. 1987b. Cosmic Background Temperature Anisotropy in a Minimal Isocurvature Model for Galaxy Formation. Astrophysical Journal Letters 315, L73–L76

Peebles, P. J. E. 2014. Discovery of the Hot Big Bang: What Happened in 1948. European Physical Journal H 39, 205–223

Peebles, P. J. E. 2017. Robert Dicke and the Naissance of Experimental Gravity Physics, 1957–1967. European Physical Journal H 42, 177–259

Peebles, P. J. E. 2020. *Cosmology's Century: An Inside History of Our Modern Understanding of the Universe*. Princeton: Princeton University Press

Peebles, P. J. E., Page, L. A., Jr., and Partridge, R. B. 2009. *Finding the Big Bang*. Cambridge: Cambridge University Press

Peebles, P. J. E., and Yu, J. T. 1970. Primeval Adiabatic Perturbation in an Expanding Universe. Astrophysical Journal 162, 815–836

Peirce, C. S. 1869. Grounds of Validity of the Laws of Logic: Further Consequences of Four Incapacities. Journal of Speculative Philosophy 2, 193–208

Peirce, C. S. 1877. The Fixation of Belief. The Popular Science Monthly 12, 1–15

Peirce, C. S. 1878a. How to Make our Ideas Clear. The Popular Science Monthly 12, 286–302

Peirce, C. S. 1878b. The Order of Nature. The Popular Science Monthly 13, 203–217

Peirce, C. S. 1892. The Doctrine of Necessity Examined. The Monist 2, 321–337

Peirce, C. S. 1903. In *Contributions to The Nation*, compiled and annotated by Kenneth Laine Ketner and James Edward Cook, Part 3, p. 127. Lubbock: Texas Tech Press, 1975

Peirce, C. S. 1907. In The Essential Peirce: Selected Philosophical Writings, Volume 2, page 399. The date, 1907, seems to be nominal; the essay went through many drafts.

Penrose, R. 1989. *The Emperor's New Mind: Concerning Computers, Minds, and the Laws of Physics*. Oxford: Oxford University Press

Penzias, A. A., and Wilson, R. W. 1965. A Measurement of Excess Antenna Temperature at 4800 Mc/s. Astrophysical Journal 142, 419–421

Perl, M. L., Feldman, G. L., Abrams, G. S., et al. 1977. Properties of the Proposed $\tau$ Charged Lepton. Physics Letters B 70, 487–490

Perlmutter, S., and Schmidt, B. P. 2003. Measuring Cosmology with Supernovae. Lecture Notes in Physics 598, 195–217

Petrosian, V., Salpeter, E., and Szekeres, P. 1967. Quasi-Stellar Objects in Universes with Non-Zero Cosmological Constant. Astrophysical Journal 147, 1222–1226

Pickering, A. 1984. *Constructing Quarks: A Sociological History of Particle Physics*. Chicago: The University of Chicago Press

Planck, M. 1900. Entropie und Temperatur strahlender Wärme. Annalen der Physik 306, 719–737

Planck Collaboration 2020. Planck 2018 results. VI. Cosmological Parameters. Astronomy and Astrophysics 641, A6, 67 pp.

Poincaré, H. 1902. *La Science et l'Hypothèse*. Paris: Flammarion

Poincaré, H. 1905. *La Valeur de la Science*, Paris: Flammarion

Popper, D. M. 1954. Red Shift in the Spectrum of 40 Eridani B. Astrophysical Journal 120, 316–321

Popper, K. M. 1935. Logik der Forschung. Vienna: Springer-Verlag

Popper, K. M. 1945. The Open Society and Its Enemies, Volume 2. London: Routledge & Sons

Popper, K. M. 1965. *Conjectures and Refutations: The Growth of Scientific Knowledge*. New York: Basic Books

Popper, K. M. 1974. In *The Philosophy of Karl Popper*, Ed. Paul Arthur Schilpp. La Salle, Illinois: Open Court Publishing Company

Pound, R. V., and Rebka, G. A. 1960. Apparent Weight of Photons. Physical Review Letters 4, 337–341

Pound, R. V., and Snider, J. L. 1964. Effect of Gravity on Nuclear Resonance. Physical Review Letters 13, 539–540

Putnam, H. 1982. Three Kinds of Scientific Realism. The Philosophical Quarterly 32, 195–200

Reasenberg, R. D., Shapiro, I. I., MacNeil, P. E., et al. 1979. Viking Relativity Experiment: Verification of Signal Retardation by Solar Gravity. Astrophysical Journal, Letters 234, L219–L221

Renn, J. 2007. *The Genesis of General Relativity: Sources and Interpretations*. Jürgen Renn, general editor. Vols. 1 and 2, Einstein's Zurich Notebook, Eds. M. Janssen, J. D. Norton, J. Renn, T. Sauer, and J. Stachel. Vols. 3 and 4, Gravitation in the Twilight of Classical Physics. Eds. J. Renn and M. Schemmel. New York, Berlin: Springer

Rey, Abel 1907. *La théorie de la physique chez les physiciens contemporains*. Paris: F. Alcan

Roberts, M. S., and Whitehurst, R. N. 1975. The Rotation Curve and Geometry of M31 at Large Galactocentric Distances. Astrophysical Journal 201, 327–346

Robertson, H. P. 1955. The Theoretical Aspects of the Nebular Redshift. Publications of the Astronomical Society of the Pacific 67, 82–98

Rubin, V. C. 2011. An Interesting Voyage. Annual Review of Astronomy and Astrophysics 49, 1–28

Rubin, V. C., and Ford, W. K., Jr. 1970a. Rotation of the Andromeda Nebula from a Spectroscopic Survey of Emission Regions. Astrophysical Journal 159, 379–403

Rubin, V. C., and Ford, W. K. 1970b. A Comparison of Dynamical Models of the Andromeda Nebula and the Galaxy. In IAU Symposium 38, The Spiral Structure of our Galaxy. Eds. W. Becker and G. I. Kontopoulos, pp. 61–68

Rudnicki, K., Dworak, T. Z., Flin, P., et al. 1973. A Catalogue of 15650 Galaxies in the Jagellonian Field. Acta Cosmologica 1, 7–164, with 36 maps

Rugh, S. E., and Zinkernagel, H. 2002. The Quantum Vacuum and the Cosmological Constant Problem. Studies in the History and Philosophy of Modern Physics 33, 663–705

Sachs, R. K., and Wolfe, A. M. 1967. Perturbations of a Cosmological Model and Angular Variations of the Microwave Background. Astrophysical Journal 147, 73–90

Sandage, A. 1961. The Ability of the 200-INCH Telescope to Discriminate between Selected World Models. Astrophysical Journal 133, 355–392

Sato, K., and Kobayashi, M. 1977. Cosmological Constraints on the Mass and the Number of Heavy Lepton Neutrinos. Progress of Theoretical Physics 58, 1775–1789

Schellenberger, G., and Reiprich, T. H. 2017. HICOSMO: Cosmology with a Complete Sample of Galaxy Clusters—II. Cosmological Results. Monthly Notices of the Royal Astronomical Society 471, 1370–1389

Schiller, F. C. S. 1910. *Riddles of the Sphinx: A Study in the Philosophy of Humanism*. New and revised edition. London: Swan Sonnenschein

Schilpp, P. A. 1949. *Albert Einstein, Philosopher-Scientist*. Evanston, Illinois: Library of Living Philosophers

Schmidt, M. 1957. The Distribution of Mass in M 31. Bulletin of the Astronomical Institutes of the Netherlands 14, 17–19

Schröter, E. H. 1956. Rotverschiebung. Der heutige Stand des Nachweises der relativistischen. Die Sterne 32, 140–150

Schwarzschild, M. 1946. On the Helium Content of the Sun. Astrophysical Journal 104, 203–207

Schwarzschild, M. 1958a. The Astrophysical Implications of Element Synthesis. In Proceedings of a Conference at the Vatican Observatory, Castel Gandolfo, May 20–28, 1957, edited by D .J. K. O'Connell. Amsterdam: North-Holland, pp 279–284

Schwarzschild, M. 1958b. *Structure and Evolution of the Stars*. Princeton: Princeton University Press

Seielstad, G. A., Sramek, R. A., and Weiler, K. W. 1970. Measurement of the Deflection of 9.602-GHz Radiation from 3C279 in the Solar Gravitational Field. Physical Review Letters 24, 1373–1376

Seldner, M., Siebers, B., Groth, E. J., and Peebles, P. J. E. 1977. New Reduction of the Lick Catalog of Galaxies. Astronomical Journal 82, 249–256

Shakeshaft, J. R., Ryle, M., Baldwin, J. E., et al. 1955. A Survey of Radio Sources Between Declinations −38° and +83°. Memoirs of the Royal Astronomical Society 67, 106–154

Shane, C. D., and Wirtanen, C. A. 1967. The Distribution of Galaxies. Publications of the Lick Observatory XXII, Part 1

Shapiro, I. I. 1964. Fourth Test of General Relativity. Physical Review Letters 13, 789–791

Shapiro, I. I., Pettengill, G. H., Ash, M. E., et al. 1968. Fourth Test of General Relativity: Preliminary Results. Physical Review Letters 20, 1265–1269

Shapley, H., and Ames, A. 1932. A Survey of the External Galaxies Brighter than the Thirteenth Magnitude. Annals of Harvard College Observatory 88, 43–75

Shostak, G. S. 1972. Aperture Synthesis Observations of Neutral Hydrogen in Three Galaxies. PhD Thesis, California Institute of Technology

Sigmund, K. 2017. *Exact Thinking in Demented Times: The Vienna Circle and the Epic Quest for the Foundations of Science*. New York: Basic Books

Smirnov, Yu. N. 1964. Hydrogen and $He^4$ Formation in the Prestellar Gamow Universe. Astronomicheskii Zhurnal 41, 1084–1089. English translation in Soviet Astronomy AJ 8, 864–867, 1965

Smith, H. E., Spinrad, H., and Smith, E. O. 1976. The Revised 3C Catalog of Radio Sources: A Review of Optical Identifications and Spectroscopy. Publications of the Astronomical Society of the Pacific 88, 621–646

Smith, S. 1936. The Mass of the Virgo Cluster. Astrophysical Journal 83, 23–30

Smoot, G. F., Bennett, C. L., Kogut, A., et al. 1992. Structure in the COBE Differential Microwave Radiometer First-Year Maps. Astrophysical Journal Letters 396, L1–L5

Soneira, R. M. and Peebles, P. J. E. 1978. A Computer Model Universe: Simulation of the Nature of the Galaxy Distribution in the Lick Catalog. Astronomical Journal 83, 845–861

Stachel, J., Klein, M. J., Schulman, R., et al. Eds. 1987. *The Collected Papers of Albert Einstein*; English companion volumes translated by A. Beck, A. Engel, A Hentschel. Princeton: Princeton University Press

Steigman, G., Sarazin, C. L., Quintana, H., and Faulkner, J. 1978. Dynamical Interactions and Astrophysical Effects of Stable Heavy Neutrinos. Astronomical Journal 83, 1050–1061

St. John, C. E. 1928. Evidence for the Gravitational Displacement of Lines in the Solar Spectrum Predicted by Einstein's Theory. Astrophysical Journal 67, 195–239

St. John, C. E. 1932. Observational Basis of General Relativity. Publications of the Astronomical Society of the Pacific 44, 277–295

Sunyaev, R. A. 2009. When We Were Young ... In Finding the Big Bang. Eds. P. J. E. Peebles, L. A. Page, Jr., and R. B. Partridge. Cambridge: Cambridge University Press, pp. 86–92

Tolman, R. C. 1934. *Relativity, Thermodynamics, and Cosmology*. Oxford: Clarendon Press

Tonry, J. L., Schmidt, B. P., Barris, B., et al. 2003. Cosmological Results from High-z Supernovae. Astrophysical Journal 594, 1–24

Trimble, V., and Greenstein, J. L. 1972. The Einstein Redshift in White Dwarfs. III. Astrophysical Journal 177, 441–452

Trumpler, R. J. 1956. Observational Results on the Light Deflection and on Red-shift in Star Spectra. In Jubilee of Relativity Theory. Eds. André Mercier and Michel Kervaire. Helvetica Physica Acta, Suppl. IV, pp. 106–113

Turner, M. S., Wilczek, F., and Zee, A. 1983. Formation of Structure in an Axion-Dominated Universe. Physics Letters B 125, 35–40

van de Hulst, H. C., Raimond, E., and van Woerden, H. 1957. Rotation and Density Distribution of the Andromeda Nebula Derived from Observations of the 21-cm Line. Bulletin of the Astronomical Institutes of the Netherlands 14, 1–16

Vessot, R. F. C., and Levine, M. W. 1979. A Test of the Equivalence Principle Using a Space-Borne Clock. General Relativity and Gravitation 10, 181–204

Vessot, R. F. C., Levine, M. W., Mattison, F. M., et al. 1980. Test of Relativistic Gravitation with a Space-Borne Hydrogen Maser. Physical Review Letters 45, 2081–2084

von Weizsäcker, C. F. 1938. Über Elementumwandlungen im Innern der Sterne. II. Physikalische Zeitschrift 39, 633–646

Vysotskiĭ, M. I., Dolgov, A. D., and Zel'dovich, Y. B. 1977. Cosmological Limits on the Masses of Neutral Leptons. Zhurnal Eksperimental'noi i Teoreticheskoi Fiziki Pis'ma 26, 200-202. English translation in Journal of Experimental and Theoretical Physics Letters 26, 188–190

Weinberg, S. 1989. The Cosmological Constant Problem. Reviews of Modern Physics 61, 1–23

Weinberg, S. 1992. Dreams of a Final Theory. New York: Pantheon

Weinberg, S. 1996. Sokal's Hoax. New York Review of Books 43, no. 13, 11–15

Wheeler, J. A. 1957. The Present Position of Classical Relativity Theory, and Some of Its Problems. In Proceedings of the Conference on the Role of Gravitation in Physics: At the University of North Carolina, Chapel Hill, January 18–23, 1957, pp. 1–5. Eds. Cécile DeWitt and Bryce DeWitt.

White, S. D. M., and Rees, M. J. 1978. Core Condensation in Heavy Halos: A Two-Stage Theory for Galaxy Formation and Clustering. Monthly Notices of the Royal Astronomical Society 183, 341–358

Wigner, E. P. 1960. The Unreasonable Effectiveness of Mathematics in the Natural Sciences. Communications in Pure and Applied Mathematics 13, 1–14

Wilczek, F. A. 2015. A Beautiful Question: Finding Nature's Deep Design. New York: Penguin Press

Wildhack, W. A. 1940. The Proton-Deuteron Transformation as a Source of Energy in Dense Stars. Physical Review 57, 81–86

Will, C. M. 2014. The Confrontation between General Relativity and Experiment. Living Reviews in Relativity 17, 117 pp.

Zel'dovich, Ya. B. 1962. Prestellar State of Matter. Zhurnal Eksperimental'noi i Teoreticheskoi Fiziki 43, 1561–1562. English translation in Soviet Journal of Experimental and Theoretical Physics 16, 1102–1103, 1963

Zel'dovich Y. B., 1968. The Cosmological Constant and the Theory of Elementary Particles. Uspekhi Fizicheskikh Nauk 95, 209–230. English translation in Soviet Physics Uspekhi 11, 381–393

Zel'dovich, Y. B. 1972. A Hypothesis, Unifying the Structure and the Entropy of the Universe. Monthly Notices of the Royal Astronomical Society 160, 1P–3P

Zuckerman, H. A. 2016. The Sociology of Science and the Garfield Effect: Happy
    Accidents, Unanticipated Developments and Unexploited Potentials. Frontiers in
    Research Metrics and Analytics 3, Article 20, 19 pp.

Zwicky, F. 1929. On the Red Shift of Spectral Lines through Interstellar Space.
    Proceedings of the National Academy of Science 15, 773–779

Zwicky, F. 1933. Die Rotverschiebung von Extragalaktschen Nebeln. Helvetica Phys-
    ica Acta 6, 110–127

Zwicky, F. 1937. On the Masses of Nebulae and of Clusters of Nebulae. Astrophysical
    Journal 86, 217–246

Zwicky, F., Herzog, E., Wild, P., Karpowicz, M., and Kowal, C. T. 1961–1968. *Cat-
    alogue of Galaxies and Clusters of Galaxies*, in 6 volumes. Pasadena: California
    Institute of Technology

# A NOTE ON THE TYPE

THIS BOOK HAS been composed in Miller, a Scotch Roman typeface designed by Matthew Carter and first released by Font Bureau in 1997. It resembles Monticello, the typeface developed for The Papers of Thomas Jefferson in the 1940s by C. H. Griffith and P. J. Conkwright and reinterpreted in digital form by Carter in 2003.

Pleasant Jefferson ("P. J.") Conkwright (1905–1986) was Typographer at Princeton University Press from 1939 to 1970. He was an acclaimed book designer and aiga Medalist.

The ornament used throughout this book was designed by Pierre Simon Fournier (1712–1768) and was a favorite of Conkwright's, used in his design of the *Princeton University Library Chronicle*.